Advanced Extraction Techniques for Drug and Toxin Detection

Abdul Khader Karakka Kal ·
Michael Benedict Subhahar · Tajudheen K. Karatt

Advanced Extraction Techniques for Drug and Toxin Detection

A Practical Guide to Sample Preparation in Forensic, Clinical, and Anti-Doping Analysis

Springer

Abdul Khader Karakka Kal
Equine Forensic Unit
Central Veterinary Research Laboratory
Dubai, United Arab Emirates

Michael Benedict Subhahar
Equine Forensic Unit
Central Veterinary Research Laboratory
Dubai, United Arab Emirates

Tajudheen K. Karatt
Equine Forensic Unit
Central Veterinary Research Laboratory
Dubai, United Arab Emirates

ISBN 978-3-032-17334-8 ISBN 978-3-032-17335-5 (eBook)
https://doi.org/10.1007/978-3-032-17335-5

This Springer imprint is published by the registered company Springer Nature Switzerland AG
The registered company address is: Gewerbestrasse 11, 6330 Cham, Switzerland

Foreword

It is both an honour and a deeply personal privilege to write this foreword for *Advanced Extraction Techniques for Drug and Toxin Detection: A Practical Guide to Sample Preparation in Forensic, Clinical, and Anti-doping Analysis*, authored by my esteemed colleagues Dr. Abdul Khader Karakka Kal, Dr. Michael Benedict Subhahar, and Dr. Tajudheen K. Karatt of the Equine Forensic Unit, Central Veterinary Research Laboratory (CVRL), Dubai.

Having had the distinction of being the first employee of the Equine Forensic Unit and witnessing its remarkable evolution since its inception in 1993, I take immense pride in the progress the laboratory has achieved over the decades. From its modest beginnings, the laboratory has developed into one of the most advanced, respected, and internationally recognized centres for anti-doping and forensic drug testing. Its success has always been built upon a foundation of scientific precision, ethical integrity, and an unwavering dedication to the welfare of equine sports.

This journey of excellence would not have been possible without the visionary guidance and generous patronage of His Highness Sheikh Mohammed Bin Rashid Al Maktoum, Vice President and Prime Minister of the United Arab Emirates and Ruler of Dubai. His Highness's steadfast support for scientific advancement, innovation, and equine welfare has provided the inspiration and resources that allowed the CVRL and its specialized units to flourish and set new benchmarks in veterinary and forensic research.

Equally pivotal has been the leadership of Dr. Ali Ridha, Director General of CVRL, whose foresight, encouragement, and insistence on research excellence have elevated the institution to global prominence. Under his guidance, the laboratory has consistently upheld the highest standards of analytical capability and international accreditation.

This book reflects the depth of scientific expertise and the spirit of innovation that have defined the Equine Forensic Unit over the years. By focusing on advanced extraction techniques—the crucial first step in accurate analytical testing—the authors provide valuable insights into the methodologies that underpin modern anti-doping and forensic analysis. The text captures both the theoretical principles and practical approaches that every analytical scientist must understand to achieve precision, reliability, and reproducibility in drug testing.

Drawing from years of collective experience, the authors successfully bridge the disciplines of analytical chemistry, forensic toxicology, and anti-doping research. They present complex scientific concepts with remarkable clarity, making this volume both intellectually rigorous and practically valuable for professionals and students alike.

Advanced Extraction Techniques for Drug and Toxin Detection: A Practical Guide to Sample Preparation in Forensic, Clinical, and Anti-doping Analysis will undoubtedly serve as an essential reference for researchers, analysts, and scholars striving to advance the science of drug testing and uphold the principles of fairness and integrity in competitive sport.

Peter Albert
Former Chief Analyst
Equine Forensic Unit
Central Veterinary Research
Laboratory
Dubai, United Arab Emirates

Preface

Since its inception in 1985 under the visionary leadership of His Highness Sheikh Mohammed Bin Rashid Al-Maktoum, the Central Veterinary Research Laboratory (CVRL) has become a pivotal institution for veterinary diagnostics, research, and forensic testing in the UAE and the broader region. What began as a modest diagnostic facility with only two staff members has evolved steadily into a premier center, now employing over a hundred dedicated scientists and technical personnel across multiple specialized fields.

Growing up in Dubai, one cannot ignore the profound cultural significance of animal sports, notably horse and camel racing. While camel racing reflects our ancestral heritage, horse racing has developed into a symbol of modern prestige and competitive excellence. As these sports have risen on the global stage, so too has the imperative to safeguard their integrity, ensuring animals are free from doping and performance-enhancing substances. This mission is not only a regulatory requirement but one that protects our cultural values and the welfare of these animals.

In 1993, the establishment of the Equine Forensic Unit under the patronage of HH Sheikh Mohammed bin Rashid Al Maktoum marked a landmark achievement. The unit was the first accredited racing chemistry laboratory in the Middle East and North Africa, tasked with rigorous testing of biological samples to uphold the standards set by the Emirates Racing Authority and related organizations. Initially, testing relied on foundational techniques such as thin-layer chromatography and colorimetric assays. However, advancements over the years have brought in state-of-the-art methodologies including high-performance liquid chromatography and mass spectrometry, significantly enhancing detection capabilities.

The book "**Advanced Extraction Techniques for Drug and Toxin Detection: A Practical Guide to Sample Preparation in Forensic, Clinical, and Anti-Doping Analysis**" arrives at a time when the landscape of anti-doping science is rapidly evolving. As new pharmacological agents emerge and detection thresholds become ever more stringent, the sophistication of analytical science must advance in parallel. Extraction remains the foundation of this process; without it, even the most advanced instrumentation cannot deliver reliable results. Through careful optimization, validation, and adherence to international standards, we can continue to uphold the credibility of forensic evidence and ensure fairness in all competitive arenas.

This book aims to share both core principles and cutting-edge techniques in sample preparation and drug detection. It addresses the evolving landscape of forensic toxicology and anti-doping science crucial for maintaining fairness in sport and public health. The book is structured into six comprehensive chapters, covering key topics from forensic toxicology's role in law and health, doping regulations, biological sample matrices, to detailed extraction methodologies, and comparative case studies on detection methods. This work reflects a commitment to innovation and excellence, inspired by the combined efforts of scientific communities, regulatory bodies, and dedicated practitioners.

The progress reflected in these pages is not the achievement of one individual but the result of decades of collective effort. I would like to extend my heartfelt congratulations and appreciation to the authors **Dr. Abdul Khader Karakka Kal, Dr. Michael Benedict Subhahar**, and **Dr. Tajudheen K. Karatt** for their dedication, expertise, and scholarly contribution to this important field. This book will undoubtedly serve as a valuable reference for students, forensic scientists, veterinary pathologists, and anti-doping professionals worldwide.

Finally, this volume is dedicated to all professionals trainers, veterinarians, scientists, and support staff whose commitment ensures clean sport and animal safety. It is hoped that this work will serve as a valuable resource and inspire continued progress in forensic and anti-doping science.

Dr. Ali Ridha Al Hashmi
Director General
Central Veterinary Research
Laboratory
Dubai, UAE

Acknowledgments The realization of *Advanced Extraction Techniques for Drug and Toxin Detection: A Practical Guide to Sample Preparation in Forensic, Clinical, and Anti-doping Analysis* has been made possible through the vision, guidance, and collective effort of many remarkable individuals whose dedication transformed an idea into reality.

We begin by extending our deepest gratitude to **Dr. Ali Ridha Al Hashmi**, Director General of the Central Veterinary Research Laboratory, whose visionary leadership, unwavering encouragement, and steadfast support formed the foundation of this endeavor. His enduring commitment to advancing veterinary science and forensic research continues to inspire excellence throughout the laboratory and beyond.

We extend our sincere appreciation to **Mr. Peter Albert**, Former Chief Analyst of the Equine Forensic Unit, whose early leadership and invaluable contributions were instrumental in shaping the foundation of the unit. His professionalism, technical expertise, and dedication to analytical excellence during its formative years helped establish the rigorous scientific standards that continue to define the laboratory today. Now retired, his legacy endures with deep respect and gratitude.

We wish to express our sincere thanks to **Dr. U. Wernery**, Scientific Director of the Central Veterinary Research Laboratory, for his invaluable mentorship, insightful feedback, and unwavering dedication to fostering scientific innovation. His guidance has been instrumental in strengthening the scientific rigor and integrity reflected throughout this book.

We also extend profound gratitude to **Mr. Mtashr Awadh Latef**, Director of Operations, and **Mrs. Farida Al Hashimi**, Director of the Director General's Office, for their steadfast encouragement, exceptional logistical coordination, and continuous administrative support, all of which have been essential to the successful completion of this work.

Special thanks are owed to **Dr. Fatma Mohammed Graiban Almheiri**, Senior Veterinary Scientist, Central Veterinary Research Laboratory, whose exemplary leadership as a veterinarian in overseeing drug administration and sample collection, along with her insightful advice and scientific input, significantly advanced the research presented herein.

We are equally grateful to **Dr. Jahfar Nalakath**, Chief Analyst of the Camel Forensic Unit, whose technical expertise and analytical insight contributed greatly to refining the extraction methodologies and analytical frameworks explored in this publication.

Our sincere appreciation goes to every member of the Equine Forensic Unit—**Mrs. Susan Alexander, Mr. Ramy Sayed, Mr. Moses Philip, Mr. Binoy Mathew, Mr. Samir Meissir, Mrs. Shino Ann Koshy, Mr. Zubair Perwad, Mr. Muhammed Ajeebsanu,** and **Mr. Mohammed Ali**—for their dedication, teamwork, and meticulous analytical skills. Their collaborative spirit, enthusiasm, and commitment to excellence were vital in transforming research findings into this comprehensive volume.

Finally, we extend our sincere appreciation to everyone at the Central Veterinary Research Laboratory whose direct and indirect contributions enriched this work. The synergy of vision, science, and dedication within this institution continues to drive innovation in anti-doping and forensic toxicology, upholding integrity, precision, and excellence in every analysis performed.

To all who supported, inspired, and believed in this journey—thank you. This book stands as a tribute to your collective commitment to advancing forensic science and promoting fair, ethical, and scientifically sound practices in anti-doping research.

Competing Interests The authors have no competing interests to declare that are relevant to the content of this manuscript.

Contents

About the Authors

Dr. Abdul Khader Karakka Kal is a senior researcher and Head of the Scientific Research Section at the Equine Forensic Unit, Central Veterinary Research Laboratory, Dubai. He holds a Ph.D. in Chemistry and brings nearly 20 years of experience in analytical chemistry, forensic toxicology, discovery chemistry, and anti-doping science. His work focuses on improving the detection of drugs and emerging performance-enhancing substances in biological samples.

Dr. Karakka Kal has authored over 40 peer-reviewed publications in high-impact journals, several of which have been recognized as Top Cited Articles. His research includes the development of advanced extraction methods, mass spectrometric applications, and metabolic investigations of novel performance-enhancing substances. His expertise uniquely positions him to bridge the gap between method development and real-world forensic and anti-doping applications. By combining deep technical knowledge with practical laboratory leadership, Dr. Kal brings a comprehensive perspective to advancing extraction techniques for forensic and clinical toxicology.

Dr. Michael Benedict Subhahar serves as Chief Analyst at the Equine Forensic Unit, Central Veterinary Research Laboratory, Dubai, bringing over 30 years of expertise in forensic racing chemistry. He earned his Ph.D. in Forensic Science from the University of Central Lancashire, U.K., and specializes in method development, screening, detection, and confirmatory analysis of performance-enhancing substances in horses and camels.

Dr. Subhahar manages a team analyzing more than 13,500 racehorse samples annually using advanced LC/MS, GC/MS, ICP/MS, and QTOF-MS platforms. He has been instrumental in establishing anti-doping laboratories, including the Camel Forensic Laboratory, and has led pioneering research on metabolic studies in race animals. His scholarly contributions include over 25 peer-reviewed publications in *Rapid Communications in Mass Spectrometry*, *Drug Testing and Analysis*, and *Journal of Analytical Toxicology*. A Fellow of the Association of Biomedical Scientists and a member of the Association of Official Racing Chemists, Dr. Subhahar is internationally recognized for advancing doping control in veterinary sports.

Dr. Tajudheen K. Karatt is an accomplished scientist with over 19 years of expertise in Racing Chemistry and Discovery Drug Chemistry. He currently serves as Deputy Technical Manager at the Equine Forensic Unit, Central Veterinary Research Laboratory, Dubai, where he oversees technical and scientific operations, and leads analytical method development, validation, and forensic research.

Dr. Karatt holds a Ph.D. in Chemistry from Bharathidasan University, India, and has a distinguished academic record, earning First Rank and Gold Medals in both his bachelor's and master's degrees in Industrial Chemistry. His professional journey includes significant roles at Biocon Bristol Myers Squibb Research Center in India, where he specialized in medicinal and synthetic chemistry before transitioning to forensic analytical science in the UAE. Dr. Karatt is a Professional Member of the Association of Official Racing Chemists (AORC) and has contributed to over 30 research publications in reputed international journals. His career is marked by numerous awards, international trainings, and certifications, making him a recognized leader in his field.

Abbreviations

AAFS	American Academy of Forensic Sciences
ABP	Athlete Biological Passport
ACE	Angiotensin-Converting Enzyme
ADAMS	Anti-Doping Administration and Management System
ADEL	Athlete Biological Passport Electronic Library
ADHD	Attention-Deficit Hyperactivity Disorder
AEA	Anandamide
AFLD	French Anti-Doping Agency
AI	Artificial Intelligence
AIDS	Acquired Immunodeficiency Syndrome
AOIRMS	Association of Olympic International Research in Sports Medicine
ARBs	Angiotensin Receptor Blockers
ARCI	Association of Racing Commissioners International
ATSDR	Agency for Toxic Substances and Disease Registry
BAC	Blood Alcohol Concentration
BALCO	Bay Area Laboratory Co-operative
CBC	Cannabichromene
CBD	Cannabidiol
CBG	Cannabigerol
CBN	Cannabinol
CIR-MS	Carbon Isotope Ratio Mass Spectrometry
CISM	Conseil International du Sport Militaire
CND	Commission on Narcotic Drugs
CNS	Central Nervous System
COPD	Chronic Obstructive Pulmonary Disease
COX	Cyclooxygenase
CSF	Cerebrospinal Fluid
D&I	Dilute-And-Inject
DART	Direct Analysis in Real Time
DART-MS	Direct Analysis in Real Time-Mass Spectrometry

DAT	Dopamine Transporter
DBS	Dried Blood Spots
DCOs	Doping Control Officers
DEA's	Drug Enforcement Administration's
DESI	Desorption Electrospray Ionization
DESI-MS	Desorption Electrospray Ionization Mass Spectrometry
DLLME	Dispersive Liquid-Liquid Microextraction
DMRP	Doping and Medication Review Panel
DNA	Deoxyribonucleic Acid
DPS	Dried Plasma Spots
DSS	Dried Serum Spots
DUI	Driving Under the Influence
DVB	Divinylbenzene
EADCMP	Equine Anti-Doping and Controlled Medication Programme
EBC	Exhaled Breath Condensate
EBTC	Evidence-Based Toxicology Collaboration
EDNA	Australia's Emerging Drugs Network
EMCDDA	European Monitoring Centre for Drugs and Drug Addiction
EMIT	Enzyme-Multiplied Immunoassay Techniques
EPA	Environmental Protection Agency
EPO	Erythropoietin
EPS	Extrapyramidal Symptoms
ESAs	Erythropoiesis-Stimulating Agents
ESBLs	Extended-spectrum β-lactamases
EUDA	European Union Drugs Agency
EWA	Early Warning Advisory
EWS	Early Warning Systems
FDA	Food and Drug Administration
FEI	Fédération Équestre Internationale
FTIR	Fourier transform infrared spectroscopy
GABA	γ-Aminobutyric Acid
GBGB	Greyhound Board of Great Britain
GC	Gas Chromatography
GC-MS	Gas Chromatography-Mass Spectrometry
GHB	Gamma-Hydroxybutyrate
HATS	Australia's Hunter Area Toxicology Service
hGH	Growth Hormones
HILIC	Hydrophilic Interaction Liquid Chromatography
HLB	Hydrophilic-Lipophilic Balanced
HRMS	High-Resolution Mass Spectrometry
IABRW	International Agreement on Breeding, Racing and Wagering

IACFT	International Alliance of Clinical and Forensic Toxicologists
ICP-MS	Inductively Coupled Plasma Mass Spectrometry
IEC	International Electrotechnical Commission
IFHA	International Federation of Horseracing Authorities
IFPMA	International Federation of Pharmaceutical Manufacturers and Associations
ILAC	International Laboratory Accreditation Cooperation
INCB	International Narcotics Control Board
IOC	International Olympic Committee
IRMS	Isotope Ratio Mass Spectrometry
ISO	International Organization for Standardization
ITA	International Testing Agency
ITPP	Myo-Inositol Trispyrophosphate
LADF	Laboratoire Antidopage Français
LA-ICP-MS	Laser ablation-inductively coupled plasma mass spectrometry
LAIs	Long-acting injectables
LC	Liquid Chromatography
LC-MS	Liquid Chromatography-Mass Spectrometry
LC-MS/MS	Liquid Chromatography–Tandem Mass Spectrometry
LDTD-APCI	Laser Diode Thermal Desorption Coupled with Atmospheric-Pressure Chemical Ionization
LIMS	Laboratory Information Management System
LLE	Liquid–liquid extraction
LOD	Limits of Detection
LOQ	Limits of Quantitation
MALDI	Matrix-Assisted Laser Desorption/Ionization
MDMB 4en PINACA	Methyl 3,3 dimethyl 2 (1 (pent 4 en 1 yl) 1H indazole 3-carboxamido)butanoate
MEPS	Microextraction by Packed Sorbent
MIPs	Molecularly Imprinted Polymers
MRSA	Methicillin-Resistant *Staphylococcus aureus*
MSI	Mass spectrometry imaging
MSPE	Magnetic Solid-Phase Extraction
MTBE	Methyl-Tert-Butyl Ether
NADOs	National Anti-Doping Organizations
NaOH	Sodium Hydroxide
NDEWS	National Drug Early Warning System
NET	Norepinephrine Transporter
NFLIS	National Forensic Laboratory Information System
NFLIS-MEC	National Forensic Laboratory Information System-Medical Examiner Data
NFLIS-Tox	National Forensic Laboratory Information System-Toxicology

NIDA	National Institute on Drug Abuse
NIST	National Institute of Standards and Technology
NOCs	National Olympic Committees
NPS	Novel Psychoactive Substances
NSAIDs	Non-Steroidal Anti-Inflammatory Drugs
OFN	Oxygen-Free Nitrogen
OSAC	Organization of Scientific Area Committees
OUD	Opioid-Use Disorder
PBPs	Penicillin-Binding Proteins
PDMS	Polydimethylsiloxane
PEDs	Performance-Enhancing Drugs
PESI-MS	Probe Electrospray Ionization Mass Spectrometry
PMPs	Prescription Monitoring Programs
PMR	Postmortem Redistribution
POCTs	Point-of-collection tests
POPs	persistent organic pollutants
ppb	Parts per billion
ppm	Parts per million
ppt	Parts per trillion
PTR-MS	Proton Transfer Reaction Mass Spectrometry
RADOs	Regional Anti-Doping Organizations
RFID	Radio-Frequency Identification
RMTC	Racing Medication and Testing Consortium
SARMs	Selective Androgen Receptor Modulators
SAX	Strong Anion Exchange
SBRA	Sports Betting Regulators Association
SBSE	Stir Bar Sorptive Extraction
SCX	Strong Cation Exchange
SDS	Sodium Dodecyl Sulfate
SERT	Serotonin Transporter
SHF	Société Hippique Française
SLE	Supported Liquid Extraction
SNRIs	Serotonin-Norepinephrine Reuptake Inhibitors
SOPs	Standard operating procedures
SPE	Solid-Phase Extraction
SPME	Solid-Phase Microextraction
SSRIs	Selective Serotonin Reuptake Inhibitors
SWGTOX	Scientific Working Group for Forensic Toxicology
THCV	Tetrahydrocannabivarin
TIAFT	The International Association of Forensic Toxicologists
TOF	Time-of-Flight
TUEs	Therapeutic Use Exemptions
UDS	Urine Drug Screening
UHPLC-MS/MS	Ultra-High-Performance Liquid Chromatography Coupled with Tandem Mass Spectrometry

UN-CND	United Nations Commission on Narcotic Drugs
UNESCO	United Nations Educational, Scientific and Cultural Organization
UNODC	United Nations Office on Drugs and Crime
UNODC-EWA	United Nations Office on Drugs and Crime Early Warning Advisory
US-NHANES	United States National Health and Nutrition Examination Survey
USSG	U.S. Sentencing Guidelines
US-ToxIC	United States Toxicology Investigators Consortium
VH	Vitreous Humor
VOCs	Volatile Organic Compounds
WADA	World Anti-Doping Agency
WAX	Weak Anion Exchange
WCX	Weak Cation Exchange
WHO	World Health Organization
WOAH	World Organisation for Animal Health

Chapter 1
Introduction

Abstract The continuous evolution of analytical science, coupled with the global expansion of sports and forensic applications, has reshaped the role of toxicology in both legal and regulatory frameworks. The detection of drugs, poisons, and emerging psychoactive substances has moved beyond the confines of routine laboratory testing to become a multidisciplinary field that bridges chemistry, pharmacology, forensic medicine, and public health. As the world contends with increasingly sophisticated drug design and distribution, forensic and anti-doping laboratories must rise to the challenge of detecting, interpreting, and regulating these substances with scientific rigor and ethical responsibility. The book "*Advanced Extraction Techniques for Drug and Toxin Detection: A Practical Guide to Sample Preparation in Forensic, Clinical, and Anti-Doping Analysis*" was conceived to serve as both a scientific reference and a practical guide for professionals working at the interface of toxicology, forensic investigation, and doping control. It brings together the latest developments in sample extraction, chromatographic analysis, and mass spectrometric detection methods that reinforce the integrity of forensic evidences and the credibility of sports regulation.

Keywords Forensic toxicology · Anti-doping analysis · Drug testing · Extraction techniques · Biological matrices · Sample preparation · Method validation · High-resolution mass spectrometry

1.1 The Expanding Role of Forensic and Anti-doping Toxicology

Forensic toxicology has long been the cornerstone of criminal and medico-legal investigations. It involves the scientific detection and interpretation of chemical substances in biological or environmental matrices to determine their relevance to cases of poisoning, death, or impairment. In recent years, however, the scope of the field has broadened to extend across regulatory, environmental, and therapeutic dimensions. Modern forensic toxicology contributes not only to criminal justice but

A. K. Karakka Kal et al., *Advanced Extraction Techniques for Drug and Toxin Detection*, https://doi.org/10.1007/978-3-032-17335-5_1

also to workplace safety, environmental surveillance, pharmaceutical regulation, and public-health monitoring.

In parallel, anti-doping toxicology has emerged as a specialized extension of forensic science. Rooted in the ethics of fair play, it ensures that athletes, both human and animal, compete on an equal footing, free from the influence of artificial performance enhancement. This domain has undergone profound transformation since the early twentieth century, when stimulants such as strychnine, caffeine, and amphetamines were first recognized as doping agents. The advent of anabolic steroids, peptide hormones, designer drugs, and gene-editing technologies has redefined the analytical challenges faced by modern laboratories.

The establishment of global governance bodies like the World Anti-Doping Agency (WADA) reinforced the need for robust, standardized, and scientifically defensible methods. These organizations depend heavily on forensic laboratories accredited and monitored under standards such as ISO/IEC 17,025 to deliver accurate, reproducible, and legally defensible results. Forensic toxicologists, therefore, operate at a unique intersection where science meets regulation and where analytical precision directly impacts human rights, public confidence, and animal welfare.

1.2 Evolution of Analytical Technologies

The development of powerful analytical tools has revolutionized the field of toxicology. Early forensic testing relied on colorimetric assays and thin-layer chromatography. While effective for gross screening, these methods lacked the specificity and sensitivity needed for modern doping control or forensic casework. The emergence of gas chromatography-mass spectrometry (GC–MS) in the late twentieth century marked a turning point, enabling unambiguous identification of volatile compounds and metabolites through their unique mass spectral signatures.

Subsequent advances introduced liquid chromatography–tandem mass spectrometry (LC–MS/MS), which expanded analytical capability to include thermally unstable and non-volatile molecules such as peptides, corticosteroids, and polar drugs. Today, hybrid techniques, including high-resolution mass spectrometry (HRMS), ultra-performance liquid chromatography (UPLC), and Orbitrap-based systems, allow detection at parts-per-trillion levels, transforming both forensic and anti-doping workflows.

Yet, the sophistication of these instruments does not negate the importance of the most critical preparatory step- sample extraction. Extraction determines how effectively analytes are isolated from complex matrices, purified from interfering substances, and concentrated for detection. Inadequate extraction can lead to poor recovery, matrix effects, and even false-negative results, undermining the reliability of entire investigations. Thus, extraction is not merely a preparatory step but the foundation upon which all analytical success is built.

1.3 The Central Role of Extraction in Analytical Science

Extraction techniques in forensic toxicology and anti-doping science have evolved in parallel with advancements in chromatography and mass spectrometry. Classical methods such as liquid–liquid extraction (LLE), once the mainstay of toxicological analysis, offered simplicity and accessibility but often required large solvent volumes and intensive labor. The introduction of solid-phase extraction (SPE) represented a significant leap forward, providing greater selectivity, reproducibility, and compatibility with automation.

In recent years, innovations such as solid-phase microextraction (SPME), dispersive liquid–liquid microextraction (DLLME), stir-bar sorptive extraction (SBSE), and supported liquid extraction (SLE) have reshaped the landscape of sample preparation. These miniaturized, solvent-efficient approaches not only align with the principles of green chemistry but also deliver superior recovery and enrichment factors, essential for detecting trace-level analytes in forensic and anti-doping applications.

Hair, urine, and plasma matrices each present unique challenges and opportunities. Hair, for instance, provides a retrospective timeline of drug exposure, allowing months-long detection windows. Urine remains the standard matrix for routine screening due to its non-invasive collection and high analyte concentration, while plasma offers real-time pharmacokinetic insight. Tailoring extraction protocols to the characteristics of each matrix, considering factors such as pH, polarity, protein binding, and enzymatic stability, is crucial to analytical accuracy.

1.4 Integration of Science, Law, and Public Health

The integration of scientific precision with legal accountability lies at the heart of forensic toxicology. Every result reported by a forensic or anti-doping laboratory carries potential legal or regulatory consequences. Maintaining chain-of-custody, adhering to validated methods, and complying with international standards such as SWGTOX and WADA technical documents are essential to ensuring data integrity and admissibility in court or disciplinary proceedings.

Beyond the courtroom, forensic toxicology supports broader societal objectives. It contributes to public-health surveillance by identifying emerging drug trends, monitoring the spread of synthetic opioids, benzodiazepines, or cannabinoids, and informing early-warning systems. It also aids in occupational safety assessments and environmental contamination studies. Anti-doping laboratories, for their part, support the ethical and medical protection of athletes and racing animals by detecting harmful misuse of pharmacological agents.

Through this intersection of law, medicine, and public health, forensic toxicology demonstrates its dual character: as both a reactive discipline that explains events after they occur and a proactive one that safeguards society from chemical threats before harm is done.

1.5 Doping in Sport: From Historical Practice to Modern Governance

The use of performance-enhancing substances in sport is not new. Historical accounts describe athletes in ancient Greece consuming herbal tonics and mushrooms believed to improve strength and endurance. By the early twentieth century, synthetic stimulants and narcotics had entered the athletic arena, culminating in high-profile doping fatalities that spurred early regulation. The death of cyclist Knut Jensen during the 1960 Rome Olympics is often cited as a catalyst for formal anti-doping policies.

In animal racing, the misuse of stimulants and sedatives paralleled human practices. Substances such as caffeine, theophylline, and phenylbutazone were once administered to enhance performance or mask pain, undermining both fairness and animal welfare. The establishment of the Equine Forensic Unit at the Central Veterinary Research Laboratory (CVRL), Dubai, in 1993 marked a pivotal moment in the Middle East's scientific commitment to clean sport. It became the first accredited racing chemistry laboratory in the region, aligning the UAE with global standards of analytical excellence.

Today, anti-doping governance is defined by collaborative international frameworks. The World Anti-Doping Agency (WADA) sets global policies through the World Anti-Doping Code, while organizations such as the Association of Racing Commissioners International (ARCI) and Federation Équestre Internationale (FEI) govern specific sporting domains. Together, these frameworks ensure harmonization in testing protocols, laboratory accreditation, and sanctioning procedures.

However, the challenge of doping persists, driven by the emergence of microdosing techniques, designer drugs, and gene-editing technologies that evade traditional detection. To address these, innovations such as the Athlete Biological Passport (ABP) and longitudinal profiling of biomarkers have become integral tools in maintaining fairness and deterrence.

1.6 Biological Matrices: Expanding the Horizons of Detection

The biological matrices used in forensic and anti-doping testing serve as the physical bridge between the event of drug intake and the analytical evidence derived from it. Each matrix offers a different temporal and biochemical perspective. Blood reveals immediate pharmacodynamic effects, urine captures excreted metabolites over extended windows, and hair records chronic or retrospective exposure.

Alternative matrices such as saliva, sweat, nails, and even meconium are gaining traction for specialized applications. In veterinary and equine toxicology, matrices such as vitreous humor, bile, and tissue samples provide valuable insights in postmortem or retrospective investigations. The choice of matrix directly affects method design, extraction strategy, and interpretation, demanding a nuanced understanding of pharmacokinetics and matrix effects.

Advances in micro-sampling and high-resolution mass spectrometry now permit near real time, high-throughput analysis of these diverse matrices. These innovations expand the forensic toolkit and strengthen the evidentiary foundation for regulatory enforcement and health protection alike.

1.7 Method Validation and Quality Assurance

No analytical method, however advanced, is complete without rigorous validation. Validation ensures that every procedure from extraction to detection meets predefined performance criteria for selectivity, sensitivity, linearity, precision, accuracy, recovery, and matrix effect. In the forensic and anti-doping domains, these parameters must align with international standards established by bodies such as SWGTOX, WADA, and ISO/IEC 17,025.

Through validation, laboratories demonstrate that their results are not only scientifically accurate but also legally defensible. In doping control, where analytical findings can lead to career-defining sanctions or exonerations, this rigor is indispensable. Continuous proficiency testing, instrument calibration, and internal quality audits sustain this credibility across laboratories worldwide.

1.8 The Rationale for This Book

The idea behind *Advanced Extraction Techniques for Drug and Toxin Detection: A Practical Guide to Sample Preparation in Forensic, Clinical, and Anti-Doping Analysis* emerged from the recognition that extraction remains the most crucial yet under-documented stage of toxicological analysis. While numerous publications discuss instrumental advancements, comparatively few provide an integrated, comparative exploration of extraction methodologies across multiple biological matrices and analytical platforms.

This book seeks to fill that gap by combining theoretical insights with practical case studies derived from contemporary forensic and anti-doping laboratories. By contextualizing extraction within the broader workflow of LC–MS and GC–MS analysis, it bridges the divide between academic research and applied science.

Each chapter builds upon the previous, guiding readers from foundational principles of forensic toxicology to advanced case-based discussions of extraction performance. This chapter situates toxicology at the intersection of science, law, and public health; Chap. 2 explores the historical and regulatory evolution of doping; Chap. 3 details the global prohibited framework; Chap. 4 introduces the nature and handling of biological matrices; Chap. 5 dissects conventional and modern extraction methods; and Chap. 6 presents comparative validation data for LC–MS and GC–MS detection.

1.9 The Global Context and Future Outlook

The globalization of sport, trade, and healthcare has made the control of drug misuse a transnational issue. Laboratories in one country often analyze samples collected across continents, requiring harmonized methods and interoperable standards. International cooperation through data-sharing networks, reference material exchange, and collaborative research ensures that anti-doping and forensic practices remain consistent worldwide.

Looking ahead, the integration of omics technologies, artificial intelligence (AI), and machine learning promises to redefine how data from toxicological testing is interpreted. Predictive modeling can help anticipate emerging doping trends, while automation and robotics will further streamline extraction and analytical workflows. At the same time, sustainability concerns are prompting the adoption of green analytical chemistry, reducing solvent waste, energy use, and environmental impact without compromising analytical performance.

1.10 Conclusion

"*Advanced Extraction Techniques for Drug and Toxin Detection: A Practical Guide to Sample Preparation in Forensic, Clinical, and Anti-Doping Analysis*" represents a product of decades of scientific progress, regulatory evolution, and technological innovation. It underscores the central role of extraction in analytical toxicology and its broader contribution to law, sport, and public health. In an era of rapid chemical innovation and increasing ethical scrutiny, forensic and anti-doping laboratories must uphold the highest standards of precision, transparency, and accountability.

This book aims to serve not only as a technical manual but also as an intellectual bridge connecting scientific methodology with the societal imperatives of justice, health, and fair competition. Through a comprehensive exploration of extraction techniques, biological matrices, validation parameters, and comparative case studies, it aspires to equip the next generation of toxicologists, forensic chemists, and regulatory scientists with the knowledge and critical insight necessary to meet the challenges of a complex and evolving world.

Chapter 2
Forensic Toxicology: Bridging Science, Law, and Public Health

Abstract This chapter explores forensic toxicology as a multidisciplinary science that integrates chemistry, medicine, law, and public health to address the identification, interpretation, and management of toxic substances. It provides a comprehensive overview of the discipline, emphasizing its pivotal role in criminal investigations, public-health protection, environmental monitoring, and regulatory enforcement. The chapter discusses how toxicologists detect and quantify drugs, poisons, and emerging psychoactive substances in biological and environmental matrices, ensuring accuracy, reliability, and legal defensibility. Significant advances in analytical techniques, including gas chromatography-mass spectrometry (GC–MS) and liquid chromatography-tandem mass spectrometry (LC–MS/MS), have enhanced the detection of conventional and novel compounds, supporting investigations into death, poisoning, road traffic incidents, workplace exposures, and drug-facilitated crimes. Beyond forensic casework, the chapter highlights forensic toxicology's contribution to monitoring global drug trends, guiding early-warning systems, supporting addiction treatment programs, and shaping policies for emerging high-risk substances such as synthetic opioids and benzodiazepines. The chapter also underscores the importance of legal rigor, detailing how strict chain-of-custody procedures, laboratory accreditation, and expert testimony ensure toxicological evidence withstands judicial scrutiny while safeguarding against wrongful convictions. In addition, it addresses environmental and public-health applications, including chemical incident response, pollution monitoring, and occupational safety assessments. Finally, the chapter examines international collaborations, professional networks, and standardization efforts that promote method validation, technology transfer, and training, enabling global preparedness for evolving chemical threats. By linking scientific rigor with legal accountability and public-health imperatives, forensic toxicology emerges as both a critical tool for justice and a cornerstone for safeguarding health and safety. This chapter, therefore, provides readers with an integrated understanding of the science, applications, and governance of modern forensic toxicology.

Keywords Forensic toxicology · Public health surveillance · Toxicovigilance · Novel psychoactive substances · Early warning systems · Chain of custody ·

A. K. Karakka Kal et al., *Advanced Extraction Techniques for Drug and Toxin Detection*, https://doi.org/10.1007/978-3-032-17335-5_2

International collaboration in toxicology · International accreditation · Method validation

2.1 The Expanding Role of Forensic Toxicology

Forensic toxicology plays a vital role in modern justice systems, public health, and environmental protection by detecting and interpreting toxic substances within legal and medical frameworks [1]. It represents a discipline that unites science, law, and health, making it indispensable on a global scale. By identifying harmful exposures and toxic threats, forensic toxicology not only ensures fairness in legal proceedings but also contributes to safeguarding human health and guiding policy development.

As society faces increasingly complex chemical exposures, the importance of forensic toxicology in promoting accountability and resilience continues to expand. Its applications extend well beyond crime scene investigations, offering significant contributions to human survival and environmental safety. Ultimately, this field remains a cornerstone in protecting present and future generations. When considering its relevance beyond crime scene investigations, the field of forensic toxicology has the potential to positively affect human survival in many ways such as:

- Identifying Cause of Death and Injury.
- Combatting Drug Abuse and Addiction.
- Enhancing Legal Accountability and Justice.
- Protecting Public Health and Environmental Safety.
- Advancing Scientific Innovation and Global Collaboration.

2.1.1 Identifying Cause of Death and Injury

Forensic toxicology is an essential discipline that connects scientific analysis with the legal system by detecting and interpreting toxins, drugs, and other chemicals in biological specimens. In cases of injury or death, it plays a pivotal role in determining whether such substances contributed to or directly caused harm. The insights gained not only support legal decision-making but also inform broader public-health responses and preventive measures.

Identifying Lethal Substances: Forensic toxicology relies on advanced analytical techniques such as GC–MS and LC–MS/MS to detect and measure drugs, poisons, and other chemicals in biological samples, including blood, urine, vitreous humor, tissues, and gastric contents [2].

- Prescription medications (e.g., antidepressants, opioids, benzodiazepines).
- Illicit drugs (e.g., methamphetamine, heroin, cocaine).
- Environmental chemicals and poisons (e.g., organophosphates, cyanide, heavy metals).

- Volatile substances (e.g., inhalants, ethanol, methanol).

Through systematic screening of different biological matrices such as blood, vitreous humor, and liver, forensic toxicologists can detect toxins that have accumulated to lethal levels or uncover exposures that may have aggravated pre-existing conditions.

Determining Cause and Manner of Death: Toxicological findings are integrated with autopsy and scene evidence to distinguish [3].

- The Cause of death: the immediate physiological dysfunction (i.e., respiratory depression due to opioid overdose).
- The Manner of death: classification as natural, accidental, suicidal, homicidal, or undetermined.

For example, if a postmortem examination reveals a deadly amount of tricyclic antidepressants along with evidence of heart disease, it may indicate suicide from an intentional overdose. On the other hand, finding small amounts of cyanide together with signs of smoke inhalation could suggest an accidental death caused by a fire.

Addressing Postmortem Redistribution: Following death, certain drugs may move from organs into the bloodstream, leading to artificially increased concentrations, a phenomenon known as postmortem redistribution (PMR) [4, 5]. This effect is especially common with lipophilic drugs that have a large volume of distribution, such as tricyclic antidepressants and amphetamines. To reduce the risk of misinterpretation, forensic toxicologists employ strategies such as:

- Sampling central (heart) v. peripheral (femoral) blood.
- Considering tissue-to-blood concentration gradients.
- Applying established PMR correction factors.

These practices ensure toxic levels reported reflect those at or near the time of death, rather than postmortem artifacts.

Investigation of Injuries and Impairment: In addition to investigating deaths, forensic toxicology also helps clarify the involvement of substances in non-fatal injuries [6, 7].

- Traffic accidents: assessing whether alcohol or drugs impaired a driver's ability.
- Workplace incidents: detecting solvent exposure or substance misuse that may have contributed to equipment-related accidents.
- Assaults and poisonings: identifying incapacitating substances, such as gamma-hydroxybutyrate (GHB) or other "date rape" drugs, in biological samples from victims.

In personal injury cases, toxicologists evaluate dose–response relationships and potential drug interactions to determine the extent of impairment and provide evidence of causation in both civil and criminal contexts.

Specialized Poisoning Cases: When uncommon or rare poisons are suspected such as plant alkaloids, industrial chemicals, or nerve agents toxicologists design or adapt specialized methods capable of detecting these substances at very low concentrations [8]. Historical milestones, like the nineteenth-century Stas-Otto extraction for nicotine, demonstrate how advances in analytical techniques have made it possible to identify toxins that were once undetectable, ultimately aiding in the resolution of complex homicide cases.

Legal and Public Health Implications: Toxicological findings play a crucial role in multiple domains, including:

- Coroners' inquests and death certification: ensuring mortality records are accurate and reliable.
- Criminal prosecutions: offering scientifically validated evidence of intoxication or poisoning.
- Epidemiological monitoring: identifying overdose patterns and detecting emerging toxic threats.
- Policy and regulation: guiding decisions on drug scheduling, workplace safety standards, and environmental protections.

Forensic toxicology remains vital in clarifying how substances contribute to both deaths and injuries. Through the integration of advanced technologies, cross-disciplinary expertise, and careful interpretation, it provides accurate conclusions that strengthen justice systems while supporting public-health initiatives. As the drug landscape continues to evolve, sustained innovation and harmonized standards are essential to meet future challenges.

2.1.2 Combatting Drug Abuse and Addiction

Forensic toxicology the examination of biological specimens and seized materials for drugs and toxins plays a pivotal role in addressing drug abuse and its consequences. By analyzing blood, urine, tissues, and illicit substances, forensic laboratories generate essential data that highlight which drugs are most prevalent, which are newly emerging, and which pose significant harm [7, 9–14]. When aggregated, these findings reveal patterns of drug use that can guide prevention and intervention strategies.

For example, a large-scale study of nearly 17,800 forensic cases in Shanghai (2021) identified synthetic cannabinoids as the most frequently detected class of abused drugs [9, 11]. At the national level, programs such as the United States Drug Enforcement Administration's (DEA's) National Forensic Laboratory Information System (NFLIS) compile drug identification data from hundreds of forensic laboratories, tracking both the purity and quantity of seized substances. Such data are crucial for monitoring long-term trends, shaping law-enforcement priorities, and supporting drug scheduling decisions.

Through this data-driven approach, forensic toxicology offers critical insights into illicit drug markets and patterns of abuse. Its contributions to combating drug misuse and addiction can be broadly categorized into the following areas:

- Identifying patterns of drug abuse.
- Supporting law enforcement and criminal investigations.
- Informing healthcare and addiction treatment strategies.
- Shaping drug policy and regulatory frameworks.
- Contributing to public-health surveillance and early-warning systems.

Identifying Patterns of Drug Abuse: Forensic toxicology plays a vital role in monitoring drug trends by systematically documenting thousands of cases. Each case whether involving an impaired driver, a patient, or a postmortem examination includes records of the substances detected and their concentrations [10]. When these findings are compiled over time, they provide valuable insight into changing patterns of drug use. For instance, routine screenings in Québec, Canada, identified a sharp increase in designer benzodiazepines and synthetic opioids after new analytical methods were adopted. On a broader scale, national monitoring systems such as NFLIS-Drug compile laboratory results from law-enforcement seizures, offering a clearer picture of circulating substances and their potencies [11]. Such data enables public-health authorities and researchers to identify geographic or demographic hotspots and anticipate emerging threats. Supporting this, a Chinese study emphasized that raw forensic data serves as a scientific foundation for drug abuse prevention by reflecting regional toxicology patterns. Overall, forensic toxicology databases function as real-time surveillance tools, mapping the prevalence of drugs, detecting novel psychoactive substances (NPS), and identifying high-risk polysubstance use across communities.

Global Forensic Toxicology Trends (2015–2025): Over the past decade, forensic toxicology laboratories worldwide have reported a sharp increase in novel synthetic drugs and highly potent opioids [12]. While there is no single global database, international monitoring systems provide a valuable overview of case volumes and substance types. The United Nations Office on Drugs and Crime (UNODC) Early Warning Advisory (EWA) on NPS has documented more than 8,100 toxicology cases from 31 countries between 2013 and 2024. In the United States, forensic lab reports (NFLIS-Drug) recorded approximately 1.5 million drug identifications in 2019, declining to about 1.28 million in 2020. These reports consistently show opioids and stimulants as the most common detections. Synthetic opioids (particularly fentanyl and analogues), methamphetamine, cocaine, designer benzodiazepines, synthetic cannabinoids, and other NPS dominate casework, with emerging threats like fentanyl adulterants and xylazine contributing to notable spikes. Table 2.1 summarizes the number of drug identifications reported by U.S. forensic laboratories for selected substances in 2019 and 2020. The data highlight trends in the prevalence of commonly encountered drugs, providing insight into shifts in illicit drug use and forensic casework. These statistics are valuable for understanding emerging drug threats and for informing law enforcement and public-health strategies.

Table 2.1 Number of U.S. forensic laboratory drug identifications by selected drugs in 2019 and 2020

Drug	Drug class	2019 Reports	2020 Reports
Methamphetamine	Stimulant	417,867	377,787
Cocaine	Stimulant	209,086	153,372
Fentanyl	Synthetic opioid	98,954	117,045
Heroin	Opioid	127,641	98,077
Alprazolam	Benzodiazepine	26,635	19,810
Amphetamine	Stimulant	11,242	9,051

Note Data from the DEA's NFLIS-Drug program [11]

Forensic toxicology data provide a clear window into how global drug trends have shifted over the past decade. U.S. laboratory findings from the NFLIS mirror patterns seen worldwide [11]. In 2020, methamphetamine emerged as the most frequently identified stimulant, with nearly 378,000 reports representing about 29% of all cases far outpacing cocaine, which declined by roughly 27% compared to 2019. Synthetic opioids, particularly fentanyl and its analogues, overtook heroin in casework, reflecting broader public-health crises. Reports of fentanyl increased by nearly 18% between 2019 and 2020, while heroin detections fell by about 23%. These findings align with U.S. overdose data, where synthetic opioid deaths rose from about 50,000 in 2019 to over 73,000 in 2022. Benzodiazepines also remained prominent, though patterns shifted: alprazolam reports declined by 26% in 2020 as novel designer benzodiazepines began to dominate forensic case files [11]. Polydrug use was widespread, with data showing that more than 90% of benzodiazepine-involved overdose deaths in the U.S. also included opioids.

International data highlight similar transformations. NPS once emerged at a rapid pace, with 400–500 new compounds reported annually between 2013 and 2021. However, their rate of introduction has slowed dramatically, with just 44 new NPS identified in 2022 and 31 in 2023 [11]. Despite this decline, synthetic opioids now make up the largest proportion of new NPS (about 36% in 2023), followed by stimulants and synthetic cannabinoids at roughly 24% each. Casework increasingly detects benzodiazepine-type NPS such as bromazolam and clonazolam, often combined with fentanyl. UNODC toxicology data from 2013–2024 indicate bromazolam was present in about half of NPS-related postmortem cases, while xylazine a veterinary sedative has appeared alongside fentanyl in numerous fatalities [14].

Regional patterns further illustrate the diversity of emerging threats. In East and Southeast Asia, ketamine has become a dominant finding. Nearly 90% of drug-use cases submitted from that region in recent years involved ketamine, often used alone. Seizures have also risen sharply, with more than 23 tons confiscated in 2023 [9]. Designer benzodiazepines have proliferated in Europe and North America, with 36 novel compounds reported to the European Monitoring Centre for Drugs and Drug Addiction (EMCDDA) since 2007. Cheap and accessible, these substances frequently appear in opioid-related overdoses, driving concerns about "benzo-dope" mixtures. Synthetic cannabinoids, once the largest NPS category, have stabilized but continue to appear in toxicology reports, with compounds like MDMB-4en-PINACA linked to severe toxicity. Meanwhile, new synthetic opioids, including nitazenes such as metonitazene, are gaining ground and are considered the fastest-growing subgroup of NPS [12].

Overall, forensic toxicology shows that traditional drugs such as heroin, prescription opioids, and cocaine have given way to methamphetamine, fentanyl, and designer substances. Between 2015 and 2025, the global drug supply has become more complex and potent, with increasing overlap between opioids, stimulants, and sedatives [9, 11]. International and national monitoring systems such as UNODC's Early Warning System, EMCDDA reporting, and NFLIS consistently highlight these evolving threats, underscoring the importance of forensic data as a surveillance tool for public health and law enforcement.

Supporting Law Enforcement and Criminal Investigations: Forensic toxicology plays a vital role in supporting law enforcement and the justice system by providing accurate, scientific evidence on the presence and effects of drugs, poisons, and toxicants [15–18]. Through advanced laboratory testing, strict quality-assurance protocols, and expert interpretation, toxicologists supply reliable information that guides investigations, links suspects to crimes, and strengthens both prosecution and defense strategies. Their work ensures that legal proceedings are carried out with objectivity, efficiency, and fairness.

Analytical Identification and Substance Characterization: Toxicologists employ advanced techniques such as GC–MS, LC–MS/MS, immunoassays, and atomic absorption spectroscopy to detect and measure a wide range of substances in biological samples and seized materials [16]. These methods not only confirm the presence of controlled drugs like heroin, cocaine, and methamphetamine but also reveal metabolites (such as 6-monoacetylmorphine from heroin) and cutting agents. Such findings provide intelligence on both individual cases and broader trafficking operations.

Intelligence and Criminal Network Linkages: By chemically profiling seized drugs, toxicologists can identify impurity "fingerprints" that trace back to specific laboratories or networks [17]. This type of analysis helps law enforcement connect geographically distant seizures, map trafficking routes, and detect shifts in manufacturing techniques. The resulting intelligence supports strategic enforcement decisions and targeted resource allocation.

Guiding Investigations and Scene Reconstruction: Early toxicology findings can shape the direction of investigations. For example, detecting volatile solvents may point to hidden chemical storage, while identifying incapacitating agents such as GHB informs both suspect questioning and victim care in sexual assault cases [18, 19]. Similarly, blood alcohol or drug concentration data help reconstruct timelines of impairment in traffic crashes or workplace accidents, giving investigators a clearer picture of events.

Expert Testimony in Court: In the courtroom, forensic toxicologists act as expert witnesses, translating complex data into clear, accessible explanations for judges and juries [20]. Their testimony on drug levels, metabolism, and toxic effects is crucial in determining causality and intent in cases involving homicide, overdose, or poisoning. Courts depend on these expert assessments when deciding whether a substance caused impairment, contributed to death, or can be ruled out as a factor.

Quality Assurance and Evidence Integrity: Accredited laboratories follow strict quality-assurance standards and chain-of-custody procedures to ensure results are valid and legally defensible [21, 22]. Every step, from crime scene collection to courtroom presentation, is documented to preserve evidence integrity and prevent tampering.

Examples of forensic toxicology in action include detecting GHB and flunitrazepam in drug-facilitated sexual assaults, identifying fentanyl analogues in overdose deaths to uncover illicit distribution patterns, and analyzing pesticide residues in workplace poisonings to support regulatory action.

In these ways, forensic toxicology transforms biological and physical samples into powerful evidence, equipping law enforcement with scientific clarity and helping ensure justice is served.

Informing Healthcare and Addiction Treatment: Forensic toxicology provides essential information that directly supports clinical care and addiction treatment [23–25]. By accurately identifying substances in patients' systems, it helps guide therapeutic decisions, monitor treatment progress, and tailor interventions at both individual and population levels.

Clinical Decision-making and Overdose Management: Toxicology testing enables clinicians to identify the drugs and metabolites responsible for a patient's condition, reducing diagnostic uncertainty and improving care. In emergency settings, urine toxicology screening has been shown to influence diagnoses in over a third of cases involving altered consciousness and nearly half of psychiatric presentations [23]. Rapid point-of-care testing, often confirmed by GC–MS, allows physicians to promptly administer life-saving interventions such as naloxone for opioid overdoses or reversal agents for sedative poisoning.

Monitoring Treatment Adherence and Preventing Relapse: In addiction programs, regular urine drug screening (UDS) objectively verifies adherence to therapies like buprenorphine while detecting relapse or diversion [24]. Evidence shows that structured UDS protocols improve patient retention, reduce misuse, and strengthen long-term outcomes in opioid dependence management.

Personalized Treatment through Pharmacogenetics: By integrating genetic profiling with toxicology, clinicians can adjust drug dosing based on metabolic variations [25]. This approach personalizes treatment, ensuring effectiveness while minimizing adverse effects in managing addiction and co-occurring conditions.

Shaping Public Health Surveillance and Early Warning Systems: Forensic toxicology is central to public-health surveillance, providing the data that fuels early-warning systems, drug policy development, and addiction treatment strategies. Laboratory findings transform raw toxicological evidence into actionable intelligence, allowing authorities to detect emerging threats, assess treatment outcomes, and regulate harmful substances with scientific precision [26, 27].

Early Warning Systems and Overdose Monitoring: In the United States, the National Drug Early Warning System (NDEWS), funded by NIDA, relies heavily on toxicology data to track NPS and regional overdose spikes [26]. Laboratory-confirmed results help identify emerging risks such as xylazine-adulterated opioids, enabling health agencies to adapt treatment protocols and expand harm-reduction services quickly. Similarly, NIDA's CODE initiative applies advanced forensic testing to improve the accuracy of overdose mortality data, ensuring that prevention efforts are based on reliable evidence [27].

Evaluating Treatment Effectiveness and Recovery: Toxicology data also help evaluate the impact of addiction treatment policies. A Swedish study found that expanding access to opioid-substitution therapy was associated with reduced opioid-related deaths, demonstrating how case data can guide clinical and preventive strategies [28]. Beyond acute care, toxicology plays a role in rehabilitation. Regular monitoring provides objective evidence of abstinence, reinforcing accountability in both court-mandated and voluntary treatment programs [29]. This type of testing has been linked to lower relapse rates and improved long-term retention.

Informing Drug Policy and Regulation: At the policy level, forensic toxicology identifies which drugs are most prevalent, newly emerging, or linked to harm. Agencies use these findings to make scheduling decisions, control precursor chemicals, and shape broader public-health and law-enforcement responses [30]. In the U.S., the DEA's NFLIS aggregates millions of drug identification reports each year. With its recent expansion to include toxicology (NFLIS-Tox) and medical examiner data (NFLIS-MEC), regulators now have timelier insight into postmortem and clinical cases, strengthening decisions on scheduling NPS and analogues of concern [11].

Legislative and International Frameworks: Toxicology findings directly influence national legislation. Under the Controlled Substances Act, the DEA considers forensic laboratory evidence when recommending scheduling actions to the Department of Health and Human Services. For instance, rising ethylphenidate seizures documented by NFLIS supported its classification as a Schedule I substance in 2023. At the international level, the United Nations Commission on Narcotic Drugs (UN-CND) uses toxicology reports alongside World Health Organization (WHO) and International Narcotics Control Board (INCB) recommendations to update global schedules. During its 68th session in 2025, delegates debated controls on synthetic opioids and designer benzodiazepines, drawing heavily on shared forensic data [31].

Precursor Control and Chemical Regulation: Forensic laboratories also analyze precursor chemicals and synthesis by-products to detect illicit manufacturing activity.

These data inform updates to national and UN precursor control lists under the 1988 Convention, helping restrict substances used to produce opioids, amphetamines, and NPS while preserving legitimate industrial uses. Such controls directly reduce clandestine production capacity.

Regional and European Perspectives: In Europe, the EMCDDA transitioning into the European Union Drugs Agency (EUDA) has developed an Early Warning System that integrates toxicology reports from member states. These findings support risk assessments and rapid-response actions, including temporary bans and harmonized monitoring strategies [32, 33]. Since 2024, the EUDA has also coordinated a European network of forensic and toxicological laboratories, further embedding laboratory science into policymaking.

Public Health Integration: Toxicovigilance data are increasingly used to inform harm-reduction programs. In the U.S., forensic findings contributed to CDC fentanyl advisories, prompting expanded naloxone distribution and targeted outreach in overdose hotspots [34]. Globally, UNODC and WHO rely on toxicology evidence when issuing treatment guidelines for new substances or revising opioid-substitution policies.

Public Health Surveillance and Early Warning Systems: Forensic toxicology plays a central role in public-health surveillance by identifying and quantifying drugs and toxicants in clinical and postmortem samples. These findings are shared with multidisciplinary early-warning systems (EWS), enabling the detection of emerging drug trends and health threats at the earliest stage. By identifying NPS, toxic adulterants, and unusual concentration patterns, toxicologists provide essential data for rapid risk assessments and evidence-based interventions. Global networks such as the United Nations Office on Drugs and Crime Early Warning Advisory (UNODC-EWA), the EMCDDA, the US-NDEWS, and Australia's Emerging Drugs Network (EDNA) depend heavily on these insights to guide clinical practice, inform policies, and strengthen community preparedness [35–37]. Collectively, these systems enhance situational awareness and support timely harm-reduction strategies.

Analytical Contributions of Forensic Toxicology: Advanced technologies such GC–MS and LC–MS/MS allow forensic laboratories to identify both established and novel drugs with precision [36]. These analyses reveal trends in polydrug use, potency variations, and the presence of toxic adulterants insights that cannot be obtained through clinical observation or surveys alone.

Integration into Early Warning Networks: EWS bring together health agencies, laboratories, law enforcement, and poison centers to exchange data in real time. Forensic laboratories contribute information on seized substances and toxicology results through standardized reporting channels, ensuring coordinated responses to emerging threats [28, 31, 35].

- UNODC-EWA: Collects data on NPS worldwide and integrates toxicology case reports through its Tox-Portal [38].

- EMCDDA-EWS: Coordinates EU member states by sharing forensic findings that underpin risk assessments and temporary controls [39].
- NDEWS (U.S.): Combines forensic results with epidemiological and wastewater monitoring to detect overdose spikes and emerging substances [40].
- EDNA (Australia): Links emergency departments and laboratories to rapidly detect harmful drugs and issue public-health alerts.

Public Health Impact and Harm Reduction: The translation of forensic data into actionable intelligence supports targeted harm-reduction measures such as naloxone distribution, festival drug-checking services, and updated overdose management guidelines. These outputs also strengthen public awareness campaigns and guide policymakers on resource allocation.

Challenges and Future Directions: Major challenges include delays in data sharing, inconsistent analytical standards, and limited funding for sustained participation. Future priorities should include integrating forensic data with syndromic and wastewater surveillance, applying predictive analytics, and fostering stronger international collaboration to improve early detection and prevention of drug-related harms.

2.1.3 Enhancing Legal Accountability and Justice

Forensic toxicology strengthens legal accountability and justice by providing reliable scientific evidence, upholding procedural safeguards, and preventing miscarriages of justice. Through rigorous analysis, standardized quality-assurance, expert testimony, and post-conviction review, toxicological data ensure that courts admit only valid evidence, defendants retain confrontation rights, and wrongful convictions are minimized [41, 42].

Ensuring Reliable Evidence and Chain of Custody: Forensic toxicology plays a pivotal role in maintaining the reliability and legal admissibility of chemical evidence. Central to this responsibility is the strict enforcement of chain-of-custody protocols, standardized documentation, accreditation requirements, and secure handling procedures [41]. These measures ensure that each specimen is collected, stored, and analyzed in a traceable manner, thereby protecting sample integrity, supporting judicial fairness, and reinforcing public confidence in forensic outcomes.

Chain-Of-Custody Documentation and Workflow: A chain-of-custody is essentially the documented timeline that tracks every stage of evidence handling, from collection to courtroom presentation [42]. In toxicology, samples such as blood, urine, or tissue are sealed in tamper-evident containers, labeled with unique identifiers, and accompanied by detailed submission forms. Upon arrival at the laboratory, each transfer is logged into a Laboratory Information Management System (LIMS) or documented on custody forms, recording dates, times, personnel, and storage locations. Continuous updates at each stage whether for analysis, aliquoting, or disposal ensure there are no gaps that could weaken the credibility of the evidence.

Standards and Accreditation: To guarantee reliability, laboratories follow international accreditation frameworks such as ISO/IEC 17,025, which mandate validated methods, strict chain-of-custody procedures, and proficiency testing. Guidelines from organizations like the Scientific Working Group for Forensic Toxicology (SWGTOX) further specify standards for personnel training, documentation, and quality control. Regular audits and external reviews confirm compliance, ensuring that toxicology results are both scientifically sound and legally defensible.

Security Measures and Sample Integrity: Protecting specimens from tampering or substitution is equally vital. Evidence bags are sealed with tamper-evident tape and stored in restricted-access facilities, often monitored by key-card systems or electronic logs. If seals must be broken for testing, containers are resealed and annotated to maintain continuity. Cross-checking digital records with physical forms adds an extra safeguard against error or unauthorized handling.

Legal Admissibility and Judicial Impact: Courts demand rigorous chain-of-custody documentation. Any break in this record can result in evidence being excluded. Landmark rulings, such as Melendez-Diaz v. Massachusetts, emphasize the importance of transparency and the defendant's right to cross-examine analysts. Clear testimony explaining custody records helps judges and juries understand the reliability of the evidence presented.

Technological Innovations: Advances such as barcode tracking, Radio-Frequency Identification (RFID) tagging, and blockchain technology now provide automated, tamper-proof custody systems. Portable field instruments, like portable GC–MS devices, allow immediate testing while maintaining secure digital logs. These innovations not only minimize human error but also streamline case management across jurisdictions.

The rigorous chain-of-custody procedures form the foundation of forensic toxicology, ensuring that each result is anchored in a transparent, verifiable, and legally defensible record of evidence handling.

Expert Witness Testimony and Confrontation Rights: Forensic toxicologists play a crucial role as expert witnesses, bridging the gap between complex laboratory data and the courtroom. Their testimony clarifies intricate analytical findings and interprets the significance of drugs, poisons, and metabolites detected in biological samples. The Sixth Amendment, as reinforced by the landmark Melendez-Diaz v. Massachusetts decision, guarantees defendants the right to confront and cross-examine these experts, ensuring that toxicological evidence remains transparent, reliable, and subject to scrutiny [42–44].

Expert Witness Role: In their courtroom capacity, forensic toxicologists explain the presence, concentration, and potential physiological effects of substances. They interpret chromatographic and spectrometric data, detail methodological constraints, and assess whether detected compounds may have contributed to impairment, injury, or death. By demystifying laboratory procedures such as sample preparation, instrument calibration, and data validation they make technical evidence accessible to judges and juries [43].

Standards for Testimony: Courts evaluate the admissibility of toxicology testimony under established standards like Daubert and Frye, focusing on scientific validity and general acceptance within the field. Accreditation programs (e.g., ISO/IEC 17,025) and guidelines from the SWGTOX ensure that only qualified analysts present evidence, maintaining high standards of reliability and reproducibility [44].

Confrontation Clause: The Supreme Court ruled that forensic analysts preparing chemical analysis certificates are considered "witnesses" under the 6th amendment and must testify in person for their results to be admitted. This ensures defendants have the opportunity to challenge the findings through direct cross-examination, rather than relying solely on written affidavits. Lower courts now require either live testimony or adherence to "notice-and-demand" statutes, giving defendants advance notice and subpoena rights.

Ensuring Reliability through Cross-examination: Cross-examination allows defense attorneys to question lab conditions, potential sources of error, interpretive decisions, and adherence to standard protocols. This scrutiny helps expose weaknesses in analytical methods, chain-of-custody documentation, or instrument calibration, reducing the risk of wrongful convictions.

Implications for Prosecution and Defense: For prosecutors, calling analysts to testify requires careful logistical planning and clear communication of technical content to a lay audience. Defendants can challenge reports through direct cross-examination or, in limited cases, via supervised surrogate witnesses, following precedents such as Bullcoming v. New Mexico and Williams v. Illinois.

Broader Significance: Requiring live analyst testimony enhances public confidence in forensic processes, reinforces due-process protections, and ensures that toxicology evidence is subjected to rigorous adversarial review. This procedural safeguard transforms laboratory data into credible, scrutinized evidence, upholding fairness and transparency in the justice system.

Preventing Wrongful Convictions and Enabling Exonerations: Forensic toxicology plays a crucial role in safeguarding justice by preventing wrongful convictions and facilitating exonerations. Its impact stems from the application of rigorously validated analytical methods, strict quality-assurance measures, standardized accreditation, confirmatory testing, and post-conviction re-analysis of preserved specimens [45]. These practices ensure that only scientifically robust evidence reaches the courtroom, reducing the likelihood of misinterpretation, cognitive bias, or reliance on outdated methods. When errors are discovered, advanced re-testing such as GC–MS for drug exposure or Deoxyribonucleic Acid (DNA) analysis of hair has overturned convictions previously based on flawed or obsolete techniques [46].

Rigorous Method Validation and Confirmatory Testing: Early drug screens often relied on presumptive colorimetric or immunoassay tests, which carried significant false-positive risks if used alone. Modern toxicology emphasizes confirmatory analyses using GC–MS or LC–MS/MS to accurately identify and quantify substances. This approach prevents innocent substances from being misclassified as illicit, safeguarding defendants against unjust outcomes.

Accreditation, Proficiency Testing, and Standardized Protocols: Laboratories accredited to ISO/IEC 17,025 adhere to documented procedures encompassing sample receipt, chain-of-custody, instrument calibration, and data verification [47]. Regular blind proficiency tests expose potential procedural errors and cognitive biases before they affect casework, enhancing reliability and credibility in judicial proceedings.

Minimizing Cognitive Bias: Research highlights that contextual bias can exacerbate forensic errors. Many laboratories now employ sequential unmasking, providing analysts only the information essential for analysis. By isolating quantitative testing from case narratives, this approach ensures interpretations remain objective and evidence based.

Post-conviction Re-analysis: Re-examining preserved biological specimens has corrected past judicial errors. Cases like Kirk L. Odom, convicted in 1982 solely on microscopic hair analysis and exonerated in 2012 via DNA testing, illustrate the transformative potential of modern toxicology. Similarly, toxicological expertise contributed to the Santae Tribble exoneration, prompting wider reviews of flawed forensic practices.

Systemic Reviews and Data-driven Exonerations: High-profile assessments of forensic methods, including FBI hair-unit reviews, have identified widespread errors and driven policy reforms. Organizations like the Innocence Project use toxicological analysis to challenge convictions where chemical evidence played a central role, revealing that unreliable forensic evidence contributes to over half of documented wrongful convictions.

By enforcing validated methodologies, robust accreditation, bias-reduction protocols, and post-conviction testing, forensic toxicology ensures that courts rely on accurate and defensible evidence, playing a pivotal role in preventing miscarriages of justice and securing rightful exonerations.

Guiding Sentencing and Proportional Justice: Forensic toxicology plays a vital role in ensuring proportional justice by providing accurate, objective measurements of a defendant's intoxication or involvement with controlled substances [48]. By quantifying drug or alcohol levels, toxicology informs sentencing decisions, helping judges calibrate penalties to match the severity of harm and culpability, maintain consistency across cases, and support individualized, fair outcomes.

Quantifying Impairment in DUI Cases: In driving under the influence (DUI) offenses, sentencing is often linked to measured blood alcohol concentration (BAC) or drug levels. Higher BAC readings typically trigger enhanced penalties: for instance, in many U.S. states, a BAC $\geq 0.15\%$ double the standard limit of 0.08% can elevate a

DUI from a misdemeanor to an aggravated offense, increasing fines and incarceration. Laboratory analyses of blood and urine provide the objective data courts rely on, reducing dependence on subjective observations and ensuring uniform application of the law.

Drug Quantity and Purity in Trafficking Cases: In drug trafficking, sentencing depends on the mass and purity of seized substances. Higher purity increases "converted drug weight," elevating offense levels under the U.S. Sentencing Guidelines (USSG). For example, 100 g of 80% pure methamphetamine equates to a converted weight of 80 g, resulting in a harsher sentence than if purity were lower. Toxicology reports therefore directly influence sentencing by providing precise measurements critical for guideline application.

Expert Interpretation and Contextualization: Beyond numerical data, toxicologists offer expert interpretation of factors such as metabolism, time since ingestion, and polydrug interactions. Two defendants with identical blood levels may have consumed substances at different times; pharmacokinetic insights help courts assess actual impairment. Similarly, reports revealing multiple substances allow judges to adjust sentences in proportion to the combined effects on harm or risk.

Promoting Consistency and Fairness: Standardized testing protocols (e.g., ISO/IEC 17,025 accreditation) ensure results are reliable and comparable across jurisdictions. This reduces disparities, supports appeals when toxicology evidence is misapplied, and enhances transparency, as methods and validations are documented in expert reports and testimony.

Balancing Public Safety and Proportionality: Forensic data help courts align sentences with both public safety and individual culpability. High-level offenders or those highly impaired may receive deterrent or incapacitative sentences, whereas individuals affected by addiction may benefit more from treatment-focused diversion. By grounding decisions in precise toxicological evidence, the justice system achieves equitable, proportionate, and evidence-based sentencing outcomes.

Deterrence, Accountability, and Public Confidence: Forensic toxicology plays a vital role in strengthening deterrence, reinforcing accountability, and sustaining public trust in the justice system [48, 49]. By providing accurate and scientifically validated evidence, it ensures that substance misuse, drug trafficking, and impaired driving are detected with high certainty. At the same time, standardized procedures and transparent practices give courts and communities confidence that every finding is reliable, defensible, and fair.

Certainty of Detection and Deterrence: Studies consistently show that the likelihood of detection is more effective at preventing crime than the severity of punishment. Forensic toxicology raises this certainty through highly sensitive tools such as GC–MS and LC–MS/MS, which can identify even trace levels of drugs or toxins. Reliable toxicology evidence in high-profile prosecutions sends a clear message that offenders face a real risk of being caught, discouraging drug-related crimes and impaired driving before they occur.

Accountability Through Standards and Accreditation: Accreditation under ISO/IEC 17,025, along with guidance from organizations like OSAC and SWGTOX, ensures laboratories follow uniform, reproducible practices [48, 49]. These standards govern all aspects of toxicological work from sample preparation and calibration to data interpretation. Proficiency testing, peer reviews, and audits further strengthen accountability by detecting errors early and enforcing corrective action, safeguarding the integrity of the evidence.

- OSAC's Registry catalogs vetted standards and best-practice guides, fostering consensus on method validation and quality assurance across jurisdictions.
- SWGTOX- the Scientific Working Group on Toxicology provides discipline-specific guidelines, further harmonizing protocols for sample preparation, instrument calibration, and data interpretation. These frameworks hold laboratories and analysts accountable: regular proficiency testing, peer reviews, and audits detect deviations and mandate corrective actions, reducing errors that could compromise justice.

Transparency and Chain of Custody: Maintaining a clear chain-of-custody is essential to protect evidentiary value. From tamper-proof packaging to electronic custody logs, every transfer and storage step is carefully documented. Such transparency allows both defense and prosecution to verify that samples remain intact and unaltered, building confidence in courtroom outcomes.

Public Trust and Ethical Oversight: Public faith in forensic science grows when laboratories openly share validation studies, error rates, and accreditation details. Recognition of institutions that adopt best practices, such as Organization of Scientific Area Committees (OSAC) standards, reinforces community confidence in the discipline. At the same time, ongoing ethical oversight ensures that emerging tools, including AI-driven analyses, are applied responsibly and without bias.

2.1.4 Protecting Public Health and Environmental Safety

Forensic toxicology stands at the intersection of science, law, and public welfare, serving as a crucial tool for protecting both human health and the environment. By detecting, identifying, and quantifying harmful substances in biological and environmental samples, toxicologists transform laboratory findings into actionable information. Their work informs regulatory standards, guides emergency responses, supports litigation against polluters, and helps communities better understand and manage risks [50]. Together, these efforts prevent disease, reduce environmental damage, and build trust in systems designed to safeguard public health and ecological integrity.

Monitoring Environmental Contaminants and Exposures: Toxicology laboratories routinely analyze soil, water, air, and living organisms to uncover pollutants such as heavy metals, persistent organic pollutants (POPs), and industrial chemicals. Using sophisticated tools like GC–MS and LC–MS/MS, they can identify toxicants at trace levels and map contamination pathways. This surveillance reveals long-lasting hazards, such as legacy pesticides, while also measuring the effectiveness of clean-up and remediation programs.

Assessing Human Exposure and Biomonitoring: By analyzing biological samples such as blood, urine, and hair, toxicologists estimate chemical exposures at both the individual and population level. Biomonitoring links internal chemical burdens to health outcomes, such as cognitive impairments caused by elevated lead levels in children. Large-scale programs like the United States National Health and Nutrition Examination Survey (US-NHANES) integrate toxicological data to highlight exposure trends, identify vulnerable populations, and guide targeted interventions.

Informing Regulatory Standards and Policies: Toxicological data directly influence public-health regulations. Agencies such as the Environmental Protection Agency (EPA) and WHO use dose–response findings to establish safe thresholds for drinking water, air quality, and soil contamination. These assessments also inform chemical registration, usage restrictions, and phase-outs, ensuring that environmental and occupational standards reflect real-world risks.

Emergency Response to Chemical Incidents: In the aftermath of chemical spills, industrial accidents, or illicit contaminations, forensic toxicologists collaborate with emergency teams to rapidly identify hazards and recommend protective measures. Agencies like the Agency for Toxic Substances and Disease Registry (ATSDR) rely on toxicological profiles and analyses to guide decontamination, medical countermeasures, and community safety decisions such as evacuation or shelter-in-place orders.

Environmental Forensics in Litigation: Forensic toxicology also plays a central role in legal accountability. Advanced techniques like isotope ratio mass spectrometry allow experts to trace pollutants whether from oil spills, pesticide misuse, or waste dumping back to their sources. Such evidence strengthens civil and criminal cases, holding industries responsible for environmental harm.

Community Health Surveillance and Education: Toxicology findings feed into health surveillance systems that identify exposure hotspots and support public outreach. Transparent reporting empowers communities to demand action, participate in policy discussions, and make informed health decisions.

By integrating precise detection, risk assessment, and public communication, forensic toxicology protects communities and ecosystems, ensuring safer environments and healthier lives.

2.1.5 Advancing Scientific Innovation and Global Collaboration

Forensic toxicology laboratories are consistently advancing analytical and interpretive techniques, employing tools such as high-resolution mass spectrometry and metabolomics to identify increasingly subtle chemical markers. At the same time, international networks and expert working groups facilitate the rapid exchange of protocols, case findings, and training resources. Such collaboration speeds up the validation of new methods, promotes consistency in quality-assurance standards, and helps integrate innovative technologies into routine practice. Collectively, these efforts not only enhance applied forensic investigations but also push the boundaries of fundamental toxicological science [51, 52].

Driving Methodological Innovation: The field of forensic toxicology is rapidly evolving through the introduction of advanced analytical tools and data-driven workflows [53, 54]. One of the most transformative technologies has been high-resolution mass spectrometry (HRMS), which allows both precise quantification of known substances and broad screening for unexpected or novel compounds. Instruments such as Orbitrap and time-of-flight (TOF) platforms deliver exact mass measurements that can distinguish closely related compounds and detect NPS at trace levels. When paired with data-independent acquisition strategies, HRMS also enables retrospective analysis of stored datasets, offering the flexibility to detect emerging drugs without reanalyzing samples. In parallel, metabolomics approaches are being explored to differentiate between acute poisoning and chronic exposure, paving the way for new biomarker panels that strengthen both clinical and medico-legal investigations.

Complementary innovations are transforming sample preparation and throughput. Techniques like microextraction by packed sorbent (MEPS) and μ-Elution Solid Phase Extraction (μ-Elution SPE) plates reduce solvent use and sample requirements, while maintaining recovery rates for diverse drug classes. These miniaturized methods not only support high-volume casework but also align with sustainability goals in modern laboratories.

Ambient ionization technologies are another breakthrough, offering rapid, minimal-preparation analysis. Methods such as Desorption Electrospray Ionization (DESI) and Direct Analysis in Real Time (DART) allow direct testing of surfaces, biological fluids, or tissues within seconds [55, 56]. Portable Direct Analysis in Real Time-Mass Spectrometry (DART-MS) systems have even been deployed for on-site screening at border crossings and crime scenes, reflecting a move toward decentralized toxicology testing. Similarly, Laser Diode Thermal Desorption coupled with atmospheric-pressure chemical ionization (LDTD-APCI) enables ultra-fast desorption and ionization, generating results in under 10 s while maintaining sensitivity in clinical and postmortem applications [57, 58].

With these high-throughput platforms generating increasingly complex data, advanced processing tools are essential. Chemometric techniques and machine learning algorithms are now used to filter noise, flag unknown features, and prioritize emerging threats, accelerating the identification process. Looking forward, portable and point-of-use mass spectrometry systems that integrate spectral libraries with compact analyzers are bringing toxicology testing closer to frontline decision-makers.

Together, these innovations from ultrahigh-resolution HRMS to solvent-free ionization, laser-based desorption, and AI-driven data interpretation are redefining forensic toxicology. They are improving sensitivity, accuracy, and speed, ensuring laboratories remain prepared to detect both established and emerging toxicants while supporting rapid, evidence-based decisions in healthcare and justice systems.

Standardization and Best-Practice Guidelines: Forensic toxicology is grounded in rigorous standardization and adherence to best-practice guidelines to guarantee that analytical findings are accurate, reproducible, and legally defensible across different laboratories and jurisdictions [59, 60]. These standardized frameworks ranging from international accreditation systems to discipline-specific method validation protocols establish clear benchmarks for laboratory competence, method reliability, and quality assurance.

Professional bodies such as the SWGTOX and the OSAC play a central role by publishing consensus guidelines on method validation, performance criteria, and interpretation of results. Their contributions ensure that emerging techniques undergo careful evaluation and validation before being accepted for widespread use. In parallel, the Evidence-Based Toxicology Collaboration (EBTC) advances transparency by promoting systematic reviews and evidence-grading systems, drawing from established practices in medicine to strengthen the credibility of toxicological assessments.

Proficiency testing and inter-laboratory comparison studies further reinforce these standards. By benchmarking new methods such as chiral separation of amphetamine enantiomers across multiple laboratories, toxicologists can verify reproducibility, detect variability, and confirm the robustness of analytical procedures in diverse settings.

Collectively, these frameworks and guidelines form the backbone of forensic toxicology. They ensure that laboratory practices remain transparent, validated, and trustworthy, ultimately safeguarding the integrity of scientific evidence presented in legal and public-health contexts.

2.2 International Accreditation and Quality Systems

2.2.1 ISO/IEC 17,025

ISO/IEC 17,025 is the cornerstone international standard that defines the requirements for laboratory competence, impartiality, and consistency in operations. It emphasizes the use of validated methods, thorough documentation, and robust quality-management systems to ensure that results remain reliable across different testing environments [61]. Accreditation under this framework requires laboratories to undergo regular audits by recognized bodies, which evaluate key aspects such as equipment calibration, personnel training, and procedural compliance. By meeting these requirements, laboratories demonstrate that their results are both accurate and reproducible, strengthening confidence in the integrity of toxicological evidence presented in legal and regulatory contexts.

2.2.2 *Organization of Scientific Area Committees (OSAC)*

Hosted by the National Institute of Standards and Technology (NIST), the OSAC plays a central role in advancing forensic science through the development and endorsement of standardized practices. OSAC maintains the Forensic Science Standards Library, where domain-specific subcommittees such as Chemistry: Seized Drugs & Toxicology produce guidelines tailored to forensic toxicology. One example is the Standard for Testing Blood in Impaired Driving Investigations, which defines minimum analytical scope, sensitivity thresholds, and reporting requirements to ensure uniformity in toxicology testing. In addition, OSAC's publicly accessible Registry provides an interactive database of approved standards, making it easier for laboratories worldwide to adopt and implement harmonized practices.

2.2.3 *Scientific Working Group for Forensic Toxicology (SWGTOX)*

The SWGTOX has been instrumental in shaping discipline-specific best practices. Its landmark 2013 publication, Standard Practices for Method Validation in Forensic Toxicology, established clear performance criteria such as selectivity, accuracy, precision, and limits of detection while also providing detailed guidance on validation study design [62]. These practices help laboratories confirm that their analytical methods are reliable, defensible, and suitable for forensic application. By offering consensus-based guidelines, SWGTOX fosters consistency across laboratories, reduces variability in results, and ensures that toxicology evidence can withstand scientific and legal scrutiny.

2.3 Method Validation and Best-Practice Guidelines

Ensuring the accuracy, reliability, and legal defensibility of toxicological analyses requires strict adherence to method validation and internationally recognized best-practice guidelines [63]. These processes confirm that laboratory techniques are not only scientifically sound but also consistently reproducible across different analysts, instruments, and jurisdictions.

2.3.1 Core Validation Parameters

Before a new or modified method is introduced into routine casework, it must undergo rigorous validation. This involves assessing key performance parameters, including linearity, bias, carry-over, matrix effects, and stability. Linearity ensures that the method produces accurate results across a wide concentration range, while bias testing confirms the absence of systematic error. Carry-over checks are essential to guarantee that analytes from one sample do not contaminate subsequent analyses. Matrix effects, often caused by interactions between analytes and complex biological fluids, are evaluated using post-extraction spiking to confirm accuracy in real-world conditions. Stability studies verify that analytes remain intact under specified storage and processing conditions. According to the SWGTOX guidelines, these validation studies must test at least five concentration levels, with each level analyzed in triplicate, to establish precision and robustness.

2.3.2 Documentation of Analytical Procedures

Validation is only effective when supported by comprehensive documentation. Standard operating procedures (SOPs) serve as the backbone of quality assurance, providing step-by-step instructions for every stage of the workflow. This includes sample receipt, labeling, storage conditions, extraction methods, instrument calibration, and data reporting protocols. Detailed SOPs ensure that results can be reproduced accurately by different analysts and provide transparency for auditors, accreditation bodies, and the courts. Well-structured documentation also protects laboratories during legal scrutiny, as it demonstrates adherence to standardized and scientifically validated methods.

2.3.3 Proficiency Testing and Inter-Laboratory Comparisons

Maintaining high standards requires continuous performance evaluation. Regular participation in blind proficiency tests and inter-laboratory comparison exercises allows laboratories to benchmark their performance against peers. Such programs, often organized by the American Academy of Forensic Sciences (AAFS) and national certification bodies, help identify systematic errors, highlight training needs, and strengthen overall reliability. Inter-laboratory comparisons also promote harmonization of methods across jurisdictions, ensuring that toxicological findings carry consistent weight regardless of where they are analyzed.

Together, rigorous validation, meticulous documentation, and ongoing proficiency testing form a comprehensive framework that safeguards the scientific and legal integrity of forensic toxicology. These best practices not only ensure accuracy and reproducibility but also build confidence among investigators, courts, and the wider public in the reliability of toxicological evidence.

2.4 Quality-Assurance Frameworks

Maintaining scientific integrity and legal defensibility in forensic toxicology relies on comprehensive quality-assurance (QA) frameworks that encompass accreditation, personnel competency, and standardized reporting practices [64, 65]. These measures collectively ensure that analytical results are accurate, reproducible, and trusted by courts, investigators, and the public.

2.4.1 Accreditation to ISO/IEC 17,025

Achieving accreditation under ISO/IEC 17,025 signifies that a laboratory operates within a fully developed quality-management system. This standard mandates rigorous documentation of procedures, corrective-action protocols to address errors, and regular internal audits to monitor compliance. Accreditation ensures that laboratory operations from sample handling to instrument calibration and data analysis are systematically controlled, fostering confidence that results are reliable and defensible across diverse forensic contexts. Periodic external assessments by accrediting bodies further validate that standards are consistently applied and maintained.

2.4.2 Personnel Competency and Training Standards

Laboratory accreditation is reinforced by ensuring that staff are highly qualified and trained to meet forensic demands. The OSAC draft Standard for Education and Training of Forensic Toxicology Personnel outlines essential qualifications for analysts and technicians. It defines responsibilities ranging from conducting analyses to reviewing data, preparing reports, and providing expert testimony in court. Continuous professional development, competency assessments, and targeted training programs ensure personnel maintain proficiency with evolving methodologies, advanced instrumentation, and emerging toxicological challenges, thereby minimizing human error and enhancing analytical accuracy.

2.4.3 Data Reporting, Review, and Interpretation

Transparent and standardized reporting is a cornerstone of quality assurance. Toxicology reports must clearly present results, associated measurement uncertainties, analytical limitations, and the complete chain-of-custody for all specimens. Such clarity ensures that legal stakeholders can interpret findings without ambiguity. Additionally, expert review processes are critical: senior analysts verify raw data, calibration records, and interpretive conclusions before finalizing reports. This multi-layered review system helps identify anomalies, confirms adherence to validated protocols, and ensures that results are robust, accurate, and legally defensible.

By integrating accreditation, personnel competency, and rigorous reporting and review mechanisms, forensic toxicology laboratories uphold the highest standards of scientific quality and integrity. These frameworks not only reinforce the credibility of analytical findings but also build public and judicial trust in the role of toxicology in investigations, court proceedings, and public safety initiatives.

2.5 International Professional Networks

Forensic toxicology has greatly benefited from international professional networks that foster collaboration, harmonize standards, and accelerate the dissemination of new methodologies [66]. By connecting practitioners across continents, these networks facilitate the rapid exchange of case studies, inter-laboratory comparisons, and guideline development, ensuring that advancements in analytical technology and best practices propagate efficiently worldwide. Such global collaboration not only enhances technical expertise but also cultivates the next generation of toxicologists through mentorship programs, travel grants, and training initiatives, strengthening international capacity to detect and respond to emerging chemical threats.

2.5.1 The International Association of Forensic Toxicologists (TIAFT)

TIAFT stands as the leading global organization for professionals in analytical and forensic toxicology [67]. The association boasts over 2,000 members spanning more than 70 countries, encompassing bench analysts, academic researchers, and law-enforcement scientists. Annual congresses, such as the upcoming 62nd Annual Meeting in Auckland, New Zealand (November 2025), alongside regional and online events, provide platforms for knowledge sharing, networking, and collaborative research. TIAFT publishes the triannual TIAFT Bulletin, which includes committee reports, methodological updates, and case notes, and offers downloadable Laboratory Guidelines for method validation and quality assurance. Additionally, the

TIAFT Web Observatory maintains an open-access repository of validated protocols and reference materials, supporting laboratories in resource-limited settings.

2.5.2 *International Alliance of Clinical and Forensic Toxicologists (IACFT)*

Established in 2020, the IACFT is a virtual, open-access network designed to bridge the clinical and forensic toxicology communities. Its multilingual online symposiums and webinars remove travel barriers, providing live captioning and recorded sessions for global accessibility. By offering free educational content and discussion forums, the IACFT empowers early-career toxicologists and professionals from low-resource regions to engage with cutting-edge topics. The alliance also hosts working groups focused on emerging drug threats, method standardization, and toxicovigilance, fostering rapid consensus and collaboration on analytical and clinical challenges.

2.5.3 *European Network of Forensic and Toxicological Laboratories (ENFSI)*

The EMCDDA, soon transitioning into the EUDA, links national forensic and toxicology laboratories across member states [31, 35]. Through its EWS, laboratories report new psychoactive substances and adverse events, enabling timely risk assessments and coordinated EU-wide control measures. The network strengthens laboratory readiness by recruiting technical experts and promoting uniform analytical capabilities. Collaborative initiatives, including joint handbooks developed with the Inter-American Drug Abuse Control Commission (CICAD), enhance cross-regional alignment in data collection, reporting, and policy frameworks.

2.6 Technology Transfer and Training

Forensic toxicology plays a pivotal role in bridging scientific innovation and practical application, ensuring that cutting-edge analytical methods are rapidly integrated into laboratories worldwide [31, 38]. Its impact extends beyond technology adoption; it also strengthens global capacity through structured training programs, professional guidelines, and ongoing competency development. By combining international capacity-building initiatives, standards-based education, professional association-led training, and academic-industry partnerships, the discipline ensures that practitioners remain adept in the face of evolving toxicological and forensic challenges.

2.6.1 *International Capacity-Building Programs*

The UNODC implements regional and national training workshops to reinforce forensic toxicology services across member states [38]. These programs cover laboratory quality assurance, method validation, and the use of field-testing kits, equipping local practitioners with the skills necessary to maintain high analytical standards. "Train-the-trainer" modules empower local experts to continue capacity-building initiatives, fostering sustainable improvements in laboratory performance. In addition, UNODC often provides equipment, reference standards, and proficiency-testing schemes, helping harmonize practices along trafficking routes and strengthening interagency cooperation.

2.6.2 *Standards-Based Education and Certification*

Professional frameworks established by the OSAC provide structured guidance for personnel training and certification [42]. Standards such as the draft "Standard for Education and Training of Forensic Toxicology Personnel" (OSAC 2021-N-0026) and the recently approved BSR/ASB 173–202 × delineate minimum qualifications, curricula, competency assessments, and continuing-education requirements for analysts in postmortem, human-performance, and general forensic toxicology roles. Placement of these standards on the OSAC registry facilitates global adoption, offering laboratories clear criteria for developing, certifying, and maintaining qualified personnel.

2.6.3 *Professional Association-Led Continuing Education*

International associations supplement formal training with ongoing education. The TIAFT maintains a "Web Observatory" of free resources, including protocols, reference materials, and case studies, and offers webinars and virtual courses through its Continuing Education Committee [66, 67]. The TIAFT Young Scientists Committee further promotes early-career development by organizing symposia, mentoring programs, and awards for practitioners under 40, fostering global collaboration and professional growth.

Similarly, the American Academy of Forensic Sciences (AAFS) provides year-round educational content, including webinars, laboratory-focused talks, and workshops at its annual meeting [63, 65]. Members of the AAFS Toxicology Section gain access to practical sessions on analytical methods, quality assurance, and human factors in forensic toxicology, along with opportunities to present research and participate in hands-on demonstrations of emerging instruments and software.

2.6.4 Academic–Industry Partnerships and Technology Transfer

Collaborations between universities, instrument manufacturers, and software developers accelerate the translation of novel methods into routine laboratory practice. These partnerships enable the implementation of high-resolution mass spectrometry workflows, miniaturized extraction techniques, and advanced chemometric algorithms. Automated data-analysis platforms integrating vendor-specific and open-source libraries streamline non-targeted screening for emerging psychoactive substances and environmental toxicants. Training modules and on-site installation support accompanying these technologies allow laboratories to adopt innovations efficiently without extensive in-house development.

2.6.5 Regional Networks and Thematic Consortia

Specialized networks enhance technology transfer and skill development on a regional scale. For example, the EDNA links hospital emergency departments with forensic laboratories, while the European Network of Forensic and Toxicological Laboratories under EMCDDA synchronizes training and proficiency-testing programs across member states [35]. These consortia provide workshops, technical guidance, and collaborative platforms to ensure consistent implementation of analytical standards. Emerging priorities include e-learning platforms with interactive simulations, blockchain-enabled tamper-proof training records, and expanded train-the-trainer programs in low-resource settings.

By integrating UNODC-led capacity building, OSAC-endorsed standards, professional association continuing education, academic-industry collaborations, and regional consortia, forensic toxicology ensures that advanced analytical methods are effectively transferred to practice. These efforts collectively sustain practitioner competency worldwide, enabling laboratories to respond swiftly and accurately to evolving toxicological threats while maintaining the highest standards of quality, reliability, and public trust.

2.7 Data Sharing and Research Consortia

Forensic toxicology has increasingly embraced data sharing and collaborative research networks to enhance analytical rigor, accelerate method validation, and respond effectively to emerging chemical threats. By aggregating results from clinical cases, postmortem analyses, and seizure investigations, these platforms facilitate trend detection, multicentre studies, and rapid dissemination of validated findings. Key initiatives include national poisoning registries, global early-warning portals, and multisite surveillance networks, all of which standardize data collection, foster inter-laboratory cooperation, and improve the comparability and impact of forensic toxicology outputs.

2.7.1 National Poisoning and Toxicology Registries

National registries provide centralized databases for toxicology and poisoning cases. For example, the U.S. Toxicology Investigators Consortium (ToxIC) compiles detailed exposure and clinical records from over 35 sites, encompassing more than 7,200 patient entries in 2022 alone. Similarly, Australia's Hunter Area Toxicology Service (HATS) collects regional poisoning data, allowing timely detection of local trends in both therapeutic and illicit exposures. On a global scale, the WHO INTOX system offers poison-control centers a standardized platform for data entry, analysis, and resource sharing, enabling cross-country comparisons of toxicological incidents. Brazil's DATATOX platform integrates laboratory, clinical, and epidemiological data to monitor toxic exposures nationally, guiding targeted interventions and policy decisions.

2.7.2 Global Early Warning and Data-Sharing Portals

Global EWS enhances real-time monitoring of NPS and other emerging toxicants. The UNODC Early Warning Advisory Tox-Portal, co-developed with TIAFT, aggregates NPS case reports from laboratories in nearly 100 countries. This platform provides registered users with real-time dashboards of substance identifications and associated harms. Biannual "Current NPS Threats" reports synthesize seizure and casework data to prioritize substances posing the greatest risk. In 2024 alone, the portal documented over 8,100 toxicology cases from 31 countries, revealing shifts in the prevalence of synthetic opioids and benzodiazepine-type NPS.

2.7.3 Multisite Surveillance and Early Warning Networks

Multisite surveillance networks integrate forensic data with epidemiological and social indicators to forecast drug trends. The US-NDEWS combines toxicology findings, emergency-department reports, law-enforcement seizures, and social-media analyses from 12 sentinel sites. Supported by NIDA, NDEWS provides near real-time data sharing with public-health stakeholders, enabling rapid alerts and informed interventions during outbreaks of novel stimulants or opioids.

2.7.4 European Laboratory Networks

European efforts, including the EMCDDA, European Network of Forensic and Toxicological Laboratories (soon EUDA), coordinate member-state labs to report NPS identifications and adverse-event data via the EMCDDA Early Warning System. By harmonizing analytical methodologies and enabling joint risk assessments, these networks strengthen regional preparedness and policy response.

2.7.5 Research Consortia and Multicenter Studies

Interdisciplinary research consortia further expand the impact of data sharing. The EBTC unites academic, regulatory, and industry partners to develop systematic-review methodologies, promoting transparency in method evaluation and data interpretation. Large multicenter initiatives, such as the Tox21 Federal Consortium and EPA ToxCast, link high-throughput bioassay data with exposure metrics, generating actionable insights for both forensic and regulatory toxicology.

Despite these achievements, challenges remain in harmonizing data formats, safeguarding privacy, and securing sustainable funding. Future priorities include adopting interoperable standards like JSON-LD, enhancing open-access platforms, and expanding consortia participation in underrepresented regions. By integrating national registries, global early-warning systems, regional surveillance networks, and interdisciplinary research collaborations, forensic toxicology has established a robust ecosystem for data sharing, accelerating scientific discovery, improving public-health interventions, and supporting evidence-based policy worldwide.

2.8 Future Directions

Forensic toxicology is entering an era defined by rapid technological innovation, expanding analytical capabilities, and increasing demands for speed, sensitivity, and environmental responsibility. Emerging HRMS and ambient-ionization techniques enable ultra-fast, non-targeted screening of known and novel toxicants, while microextraction and green-chemistry workflows reduce sample volumes and minimize solvent use. These advances not only enhance laboratory efficiency but also support sustainable practices. Simultaneously, artificial intelligence (AI), machine learning, and big-data platforms are being leveraged to improve spectral interpretation, identify unknown compounds, and facilitate interoperable data sharing. Blockchain technologies are also being explored to secure tamper-proof chain-of-custody records, strengthening both legal defensibility and public trust. Portable and point-of-care devices promise to decentralize toxicology testing, allowing rapid on-site decision-making in emergency settings, workplaces, and field operations.

Integrating metabolomics and pharmacogenomics offers unprecedented insights into individual metabolic profiles, helping interpret postmortem drug concentrations and informing personalized forensic analyses. High-resolution platforms such as Orbitrap and time-of-flight mass spectrometers, when combined with retrospective data mining, allow detection of trace-level substances, including novel psychoactive compounds that may otherwise evade routine testing. Ambient ionization methods, including DART-MS and Desorption Electrospray Ionization Mass Spectrometry (DESI-MS), enable near-instantaneous analysis of biological matrices or surfaces with minimal preparation, supporting high-throughput laboratory workflows and field deployments. Microextraction techniques like MEPS and μ-Elution SPE plates further optimize sample preparation by reducing solvent consumption by over 70% while maintaining high recovery across diverse analytes.

AI-driven data processing is increasingly applied to large spectral datasets, enhancing feature selection, automating unknown-compound annotation, and flagging emerging toxicants in non-targeted screenings. Big-data integration allows laboratories to detect trends, anticipate emerging chemical threats, and implement coordinated interventions. Portable mass spectrometers and point-of-care devices, often coupled with smartphone readouts or colorimetric sensors, are transforming real-time toxicology assessment at traffic checkpoints, emergency scenes, and bedside clinical care.

Sustainability is becoming central to analytical toxicology, with green-chemistry principles promoting solvent reduction, reagent recycling, and environmentally benign laboratory practices without compromising analytical performance. Meanwhile, global collaboration and standardization remain essential. OSAC and SWGTOX are developing standards for validation of HRMS workflows, AI-based analytics, and ambient-ionization protocols to ensure accuracy, precision, and reproducibility. International networks such as TIAFT and IACFT, along with research consortia, provide training programs and open-access protocol repositories to harmonize practice, particularly in resource-limited regions.

Looking ahead, forensic toxicology is poised to evolve from reactive casework toward proactive hazard identification, anticipating new psychoactive substances and environmental toxins before widespread harm occurs. By combining pioneering analytical methods, robust training frameworks, international standards, and collaborative data-sharing initiatives, the field can safeguard public health, environmental integrity, and justice globally, ensuring rapid, accurate, and equitable toxicological responses in an increasingly complex chemical landscape.

2.9 Summary

Forensic toxicology sits at the vital crossroads of science, law, and public health, serving as a cornerstone for justice and societal safety. Its scope extends far beyond determining the presence of substances in biological specimens; it plays a central role in clarifying causes of death and impairment, informing public-health strategies, and supporting legal processes. By integrating toxicological results with autopsy findings, scene investigations, and clinical data, forensic toxicologists provide critical insights into mortality and injury cases. Beyond postmortem analyses, the discipline contributes significantly to traffic safety investigations, workplace monitoring, and the detection of drug-facilitated crimes. Data from casework also inform broader public health and policy initiatives, including monitoring illicit drug markets, identifying NPS, and shaping national and international drug-control strategies.

The legal system relies heavily on the integrity of toxicological evidence. Rigorous chain-of-custody procedures, validated analytical methods, accredited laboratory practices, and expert testimony ensure that findings are scientifically robust and legally defensible. These measures safeguard against wrongful convictions and support equitable judicial outcomes. Forensic toxicology also extends into environmental and occupational health, detecting contaminants in human and environmental matrices to guide biomonitoring programs, regulatory standards, and interventions for chemical exposures. Increasingly, toxicologists contribute to real-time emergency responses and public-health surveillance, highlighting the field's dynamic role in proactive safety measures.

Scientific innovation continues to expand the reach and capabilities of forensic toxicology. Techniques such as high-resolution mass spectrometry, ambient ionization, and advanced microextraction methods enhance sensitivity, speed, and adaptability, allowing the detection of emerging substances at trace levels. Professional organizations and international networks facilitate global collaboration, standardize methodologies, and provide capacity building and training opportunities to ensure consistent, high-quality practice worldwide.

Ultimately, forensic toxicology does more than resolve past incidents; it anticipates emerging chemical and toxicological threats. By combining scientific precision, methodological rigor, and legal accountability, the discipline informs evidence-based drug policies, reinforces public-health initiatives, and strengthens global preparedness. Its integration of laboratory science, investigative insight, and policy guidance makes forensic toxicology indispensable in addressing the evolving challenges of modern society.

References

1. Drummer OH. Forensic toxicology. EXS. 2010;100:579–603. https://doi.org/10.1007/978-3-7643-8338-1_18.
2. de Campos EG, da Costa BRB, Dos Santos FS, et al. Alternative matrices in forensic toxicology: a critical review. Forensic Toxicol. 2022;40:1–18. https://doi.org/10.1007/s11419-021-00596-5.
3. Middleberg RA. Disposition of toxic drugs and chemicals in man. 11th edn. J Anal Toxicol. 2018;42:139. https://doi.org/10.1093/jat/bkx063
4. Pélissier-Alicot AL, Gaulier JM, Champsaur P, et al. Mechanisms underlying postmortem redistribution of drugs: a review. J Anal Toxicol. 2003;27:533–44. https://doi.org/10.1093/jat/27.8.533.

5. Drummer OH, Gerostamoulos D. Post-mortem redistribution of drugs and other factors affecting interpretation: a review. WIREs Forensic Sci. 2023;5: e1480. https://doi.org/10.1002/wfs2.1480.
6. Abdelaal GMM, Hegazy NI, Etewa RL, et al. Postmortem redistribution of drugs: a literature review. Forensic Sci Med Pathol. 2024;20:1483–90. https://doi.org/10.1007/s12024-023-00709-z.
7. Menéndez-Quintanal LM, Matey JM, del Fresno GV, et al. The state of the art in post-mortem redistribution and stability of new psychoactive substances in fatal cases: a review of the literature. Psychoactives. 2024;3:525–610. https://doi.org/10.3390/psychoactives3040033.
8. Moshiri M, Darchini-Maragheh E, Balali-Mood M. Advances in toxicology and medical treatment of chemical warfare nerve agents. Daru. 2012;20:81. https://doi.org/10.1186/2008-2231-20-81.
9. Cheng JY, Hui JW, Chan WS, et al. Interpol review of toxicology 2019–2022. Forensic Sci Int Synerg. 2023;6: 100303. https://doi.org/10.1016/j.fsisyn.2022.100303.
10. Santos IC, Maia D, Dinis-Oliveira RJ, et al. New psychoactive substances: health and legal challenges. Psychoactives. 2024;3:285–302. https://doi.org/10.3390/psychoactives3020018.
11. NFLIS-DRUG 2021 Annual Report. nflis.deadiversion.usdoj.gov/nflisdata/docs/NFLISDrug_2021AnnualReport.pdf. Accessed 1 Mar 2025.
12. Brunetti P, Giorgetti R, Tagliabracci A et al. Designer benzodiazepines: a review of toxicology and public health risks. Pharmaceuticals (Basel). 2021;14:560. https://doi.org/10.3390/ph14060560
13. Roberts A, Korona-Bailey J, Mukhopadhyay S. Notes from the field: Nitazene-related deaths—Tennessee, 2019–2021. MMWR Morb Mortal Wkly Rep 2022;71:1196–1197. https://doi.org/10.15585/mmwr.mm7137a5
14. Cano M, Daniulaityte R, Marsiglia F. Xylazine in overdose deaths and forensic drug reports in US states, 2019–2022. JAMA Netw Open. 2024;7: e2350630. https://doi.org/10.1001/jamanetworkopen.2023.50630.
15. Feng J, Wang L, Dai I, et al. Simultaneous determination of multiple drugs of abuse and relevant metabolites in urine by LC-MS-MS. J Anal Toxicol. 2007;31:359–68. https://doi.org/10.1093/jat/31.7.359.
16. Kacargil CU, Daglioglu N, Goren IE. Determination of illicit drugs in dried blood spots by LC–MS/MS method: validation and application to real samples. Chromatographia. 2020;83:885–92. https://doi.org/10.1007/s10337-020-03907-x.
17. Besacier F, Chaudron-Thozet H. Chemical profiling of illicit heroin samples. Forensic Sci Rev. 1999;11:105–119. https://portaildoc-snps.interieur.gouv.fr/index.php?lvl=notice_display&id=2382
18. Price JM, Smith RC, Miles AK, et al. Delays in blood collection and drug toxicology results among crash-involved drivers arrested for impaired driving. Traffic Inj Prev. 2024;25:667–72. https://doi.org/10.1080/15389588.2024.2333918.
19. Jones AW, Holmgren A, Ahlner J. Toxicological analysis of blood and urine samples from female victims of alleged sexual assault. Clin Toxicol (Phila). 2012;50:555–61. https://doi.org/10.3109/15563650.2012.702217.
20. O'Brien É, Nic Daeid N, Black S. Science in the court: pitfalls, challenges and solutions. Philos Trans R Soc Lond B Biol Sci. 2015;370:20150062. https://doi.org/10.1098/rstb.2015.0062.
21. Drummer OH. Good practices in forensic toxicology. Curr Pharm Des. 2017;23:5437–41. https://doi.org/10.2174/1381612823666170704123836.
22. D'Anna T, Puntarello M, Cannella G, et al. The chain of custody in the era of modern forensics: from the classic procedures for gathering evidence to the new challenges related to digital data. Healthcare. 2023;11:634. https://doi.org/10.3390/healthcare11050634.
23. Peters FT, Remane D. Aspects of matrix effects in applications of liquid chromatography-mass spectrometry to forensic and clinical toxicology—a review. Anal Bioanal Chem. 2012;403:2155–72. https://doi.org/10.1007/s00216-012-6035-2.
24. Jamshidi N, Athavale A, Tremonti C, et al. Evaluation of adherence monitoring in buprenorphine treatment: a pilot study using timed drug assays to determine accuracy of testing. Br J Clin Pharmacol. 2023;89:1938–47. https://doi.org/10.1111/bcp.15318.

25. Kazi I, Chenoweth MJ, Jutras-Aswad D, et al. Pharmacogenetics of biochemically verified abstinence in an opioid agonist therapy randomized clinical trial of methadone and buprenorphine/naloxone. Clin Pharmacol Ther. 2024;115:506–14. https://doi.org/10.1002/cpt.3112.
26. Love D, Jones NS. Interpol review of drug analysis 2019–2022. Forensic Sci Int Synerg. 2023;6: 100299. https://doi.org/10.1016/j.fsisyn.2022.100299.
27. Palamar JJ, Fitzgerald ND, Goldberger BA, et al. Monitoring illicit pentobarbital availability in the United States: a National Drug Early Warning System briefing. Drug Alcohol Depend. 2024;263: 112402. https://doi.org/10.1016/j.drugalcdep.2024.112402.
28. National Drug Early Warning System (NDEWS). National Drug Early Warning System (NDEWS) | National Institute on Drug Abuse (NIDA). https://nida.nih.gov/research-topics/trends-statistics/national-drug-early-warning-system-ndews. Accessed 14 May 2025
29. Elarabi HF, Shawky M, Mustafa N, et al. Effectiveness of incentivised adherence and abstinence monitoring in buprenorphine maintenance: a pragmatic, randomised controlled trial. Addiction. 2021;116:2398–408. https://doi.org/10.1111/add.15394.
30. Al-Matrouk A, Al-Hasan M, Naqi H, et al. Snapshot of narcotic drugs and psychoactive substances in Kuwait: analysis of illicit drugs use in Kuwait from 2015 to 2018. BMC Public Health. 2021;21:671. https://doi.org/10.1186/s12889-021-10705-z.
31. UNODC/EMCDDA reports (EWA/Early Warning Advisory). UNODC Early Warning Advisory (EWA) on New Psychoactive Substances (NPS). https://www.unodc.org/LSS/Home/NPS. Accessed 14 July 2025
32. European Union Drug Agency. A new network of forensic and toxicology laboratories to strengthen EU preparedness on drugs. https://www.euda.europa.eu/news/2024/new-network-forensic-and-toxicology-laboratories-strengthen-eu-preparedness-drugs_en. Accessed 14 July 2025
33. European Monitoring Centre for Drugs and Drug Addiction (EMCDDA). Understanding Europe's drug situation in 2024—key developments (European Drug Report 2024). https://santesecu.public.lu/dam-assets/fr/publications/r/rapport-europen-sur-les-drogues-2024/rapport-drogues-eu-2024.pdf. Accessed 14 July 2025
34. Manz LJ, Stewart A, Gladden RM et al. Detection of illegally manufactured fentanyls and carfentanil in drug overdose deaths—United States, 2021–2024. MMWR Morb Mortal Wkly Rep. 2024;73:1099–1105. https://doi.org/10.15585/mmwr.mm7348a2
35. EMCDDA. Operating guidelines for the European Union Early Warning System on new psychoactive substances. https://syntheticdrugs.unodc.org/uploads/syntheticdrugs/res/library/earlywarning_html/EMCDDA_Operating_Guidelines_for_the_EU_Early_Warning_System_on_NPS.pdf. Accessed 14 July 2025
36. Shafi A, Berry AJ, Sumnall H, et al. New psychoactive substances: a review and updates. Ther Adv Psychopharmacol. 2020;10:2045125320967197. https://doi.org/10.1177/2045125320967197.
37. Heinsvig PJ, Noble C, Dalsgaard PW, et al. Forensic drug screening by liquid chromatography hyphenated with high-resolution mass spectrometry (LC-HRMS). TrAC Trends Anal Chem. 2023;162: 117023. https://doi.org/10.1016/j.trac.2023.117023.
38. UNODC Laboratory and Scientific Service Portals. https://www.unodc.org/LSS/Page/About. Accessed 01 Oct 2025
39. EU Early Warning System. The EU Early Warning System on new psychoactive substances (NPS). https://www.euda.europa.eu/activities/eu-early-warning-system-on-nps_en. Accessed 01 Oct 2025
40. The National Drug Early Warning System (NDEWS). National Drug Early Warning System (NDEWS). https://ndews.org/. Accessed 01 Oct 2025
41. Jones S, Stypa M. Standardization in forensic toxicology—the United States experience. WIREs Forensic Sci. 2023;5: e1471. https://doi.org/10.1002/wfs2.1471.
42. OSAC Technical Guidance Documents. https://www.nist.gov/adlp/spo/organization-scientific-area-committees-forensic-science/other-work-products/osac-technical. Accessed 01 Oct 2025

43. Peters FT, Drummer OH, Musshoff F. Validation of new methods. Forensic Sci Int. 2007;165:216–24. https://doi.org/10.1016/j.forsciint.2006.05.021.
44. Scientific Working Group for Forensic Toxicology. Scientific Working Group for Forensic Toxicology (SWGTOX) standard practices for method validation in forensic toxicology. J Anal Toxicol. 2013;37:452–74. https://doi.org/10.1093/jat/bkt054.
45. Stephenson L, Van Den Heuvel C, Scott T, et al. Difficulties associated with the interpretation of postmortem toxicology. J Anal Toxicol. 2024;48:405–12. https://doi.org/10.1093/jat/bkae052.
46. Hundl C, Neuman M, Rairden A, et al. Implementation of a blind quality control program in a forensic laboratory. J Forensic Sci. 2020;65:815–22. https://doi.org/10.1111/1556-4029.14259.
47. Pierce ML, Cook LJ. Development and implementation of an effective blind proficiency testing program. J Forensic Sci. 2020;65:809–14. https://doi.org/10.1111/1556-4029.14269.
48. Wille SMR, Van Dijck K, Van Assche A, et al. The interest of a systematic toxicological analysis combined with forensic advice to improve the judicial investigation and final judgment in drug facilitated sexual assault cases. Pharmaceuticals. 2021;14:432. https://doi.org/10.3390/ph14050432.
49. Neuteboom W, Ross A, Bugeja L, et al. Quality management in forensic science: a closer inspection. Forensic Sci Int. 2024;358: 111779. https://doi.org/10.1016/j.forsciint.2023.111779.
50. Moulinec A, Arle J, Hollert H, et al. Assessing chemical pollution with biomonitoring approaches in streams and rivers: a critical review. Environ Sci Eur. 2025;37:69. https://doi.org/10.1186/s12302-025-01110-z.
51. Wille SMR, Desharnais B, Pichini S, et al. Liquid chromatography high-resolution mass spectrometry in forensic toxicology: what are the specifics of method development, validation and quality assurance for comprehensive screening approaches? Curr Pharm Des. 2022;28:1230–44. https://doi.org/10.2174/1381612828666220526152259.
52. Mosley JD, Schock TB, Beecher CW, et al. Establishing a framework for best practices for quality assurance and quality control in untargeted metabolomics. Metabolomics. 2024;20:20. https://doi.org/10.1007/s11306-023-02080-0.
53. Mogollón NGS, Quiroz-Moreno CD, Prata PS, et al. New advances in toxicological forensic analysis using mass spectrometry techniques. J Anal Methods Chem. 2018;2018:4142527. https://doi.org/10.1155/2018/4142527.
54. Maurer HH, Meyer MR. High-resolution mass spectrometry in toxicology: current status and future perspectives. Arch Toxicol. 2016;90:2161–72. https://doi.org/10.1007/s00204-016-1764 1.
55. Wójtowicz A, Wietecha-Posłuszny R. DESI-MS analysis of human fluids and tissues for forensic applications. Appl Phys A. 2019;125:312. https://doi.org/10.1007/s00339-019-2564-2.
56. Gupta S, Samal N. Application of direct analysis in real-time mass spectrometry (DART-MS) in forensic science: a comprehensive review. Egypt J Forensic Sci. 2022;12:17. https://doi.org/10.1186/s41935-022-00276-4.
57. Bynum ND, Moore KN, Grabenauer M. Evaluation of laser diode thermal desorption-tandem mass spectrometry (LDTD-MS-MS) in forensic toxicology. J Anal Toxicol. 2014;38:528–35. https://doi.org/10.1093/jat/bku084.
58. Zawada A, Rychter AM, Ratajczak AE, et al. Does gut-microbiome interaction protect against obesity and obesity-associated metabolic disorders? Microorganisms. 2021;9:18. https://doi.org/10.3390/microorganisms9010018.
59. Standard practices for method validation in forensic toxicology. American Academy of Forensic Sciences. https://www.aafs.org/asb-standard/standard-practices-method-validation-forensic-toxicology. Accessed 14 Oct 2025
60. Farhat N, Tsaioun K, Saunders-Hastings P, et al. Systematic review in evidence-based risk assessment. ALTEX. 2022;39:463–479. https://doi.org/10.14573/altex.2004111
61. ISO/IEC 17025 testing and calibration laboratories. https://www.iso.org/ISO-IEC-17025-testing-and-calibration-laboratories.html. Accessed 1 May 2025

62. Scientific Working Group for Forensic Toxicology. Standard practices for method validation in forensic toxicology. J Anal Toxicol. 2013;37:452–74. https://doi.org/10.1093/jat/bkt054.
63. Academy Standards Board. Standard practices for method validation in forensic toxicology (ANSI/ASB Standard 036, First Edition 2019). https://www.aafs.org/asb-standard/standard-practices-method-validation-forensic-toxicology. Accessed 1 May 2025
64. ISO/IEC 17025:2017. General requirements for the competence of testing and calibration laboratories. International Organization for Standardization. https://www.iso.org/standard/66912.html. Accessed 1 May 2025
65. American National Standards Institute (ANSI)/Academy Standards Board (ASB). Standard for a quality control program in forensic toxicology laboratories (ANSI/ASB Standard 054–2021). https://www.aafs.org/asb-standard/standard-quality-control-program-forensic-toxicology-laboratories. Accessed 14 Oct 2025
66. Ratansi S. International Association of Forensic Toxicologists (TIAFT). EBSCO Research Starters—Science. https://www.ebsco.com/research-starters/science/international-association-forensic-toxicologists-tiaft. Accessed 3 May 2025
67. The International Association of Forensic Toxicologists (TIAFT). https://tiaft.org/tiaft-2022-monday-highlights/. Accessed 3 May 2025

Chapter 3
Doping in Sport: History, Regulation, and Global Governance

Abstract Doping remains a critical issue in both human and animal sports, challenging the principles of fairness, safety, and public trust. Historically, the practice evolved from the use of stimulants in the nineteenth century to today's complex landscape of anabolic agents, blood manipulation, and genetic technologies. While human athletics often dominate the doping debate, animal sports including horse racing, greyhound racing, and equestrian disciplines face parallel challenges, with performance-enhancing substances posing risks not only to competition integrity but also to animal welfare. The establishment of the World Anti-Doping Agency (WADA) in 1999 provided a unified global framework for human sport through the World Anti-Doping Code, which standardized banned substance lists, testing protocols, and sanctioning procedures. Similarly, organizations such as the Association of Racing Commissioners International (ARCI), the Greyhound Board of Great Britain (GBGB), and the French Anti-Doping Agency (AFLD) have developed specialized rules and testing systems for animals, reflecting the need for robust regulation beyond human competition. Despite these frameworks, doping practices continue to evolve, with micro-dosing, designer substances, and emerging threats like gene doping challenging traditional detection. In both human and animal contexts, advances such as the Athlete Biological Passport, standardized laboratory accreditation, and international data sharing represent important tools in strengthening detection and enforcement. At the same time, education and awareness initiatives whether aimed at athletes, trainers, or animal caretakers are increasingly recognized as central to long-term prevention. This chapter explores doping across human and animal sports, emphasizing its historical roots, regulatory evolution, and governance challenges. It concludes that safeguarding integrity requires continuous innovation, international cooperation, and a balance between strict enforcement and the ethical responsibility to protect both athletes and animals.

Keywords Doping in sport · Anti-doping regulation · World anti-doping agency · Animal sports doping · Equine anti-doping · Greyhound racing · Performance-enhancing substances · Athlete biological passport

A. K. Karakka Kal et al., *Advanced Extraction Techniques for Drug and Toxin Detection*, https://doi.org/10.1007/978-3-032-17335-5_3

3.1 Definition and Historical Evolution

Doping, the use of prohibited substances or methods to enhance athletic performance threatens the integrity of sport and poses serious health risks to athletes. According to the World Anti-Doping Code, an athlete commits a doping violation by using any banned substance or method listed in Article 2. Although doping has been documented since ancient times, modern regulatory efforts emerged in response to high-profile tragedies during the mid-twentieth century, culminating in the establishment of the WADA in 1999 and the harmonized Code in 2003. Today, over 200 countries adhere to WADA's regulations, supported by comprehensive testing protocols, education initiatives, and continually evolving analytical methods.

Doping includes one or more anti-doping rule violations, including the presence of prohibited substances in an athlete's sample or the use of banned methods. Historical records reveal that athletes have long sought chemical and dietary strategies to enhance performance. Ancient Greek and Roman competitors consumed stimulants such as dried figs, herbal mixtures, hallucinogenic mushrooms, and animal-derived products to improve strength, endurance, or pain tolerance [1–5]. By the nineteenth century, endurance athletes experimented with opiate-based tonics, cocaine-brandy mixtures, and even nitroglycerine to sustain energy during prolonged competitions.

3.1.1 Ancient Origins (776 BC–Nineteenth Century)

Doping in sports dates back to ancient civilizations. Greek Olympians consumed figs, herbal mixtures, and animal testicles to boost performance, while Roman gladiators used stimulants like hallucinogenic mushrooms and wine concoctions [1, 2]. These practices aimed to enhance endurance, strength, or pain tolerance, reflecting early recognition of performance-enhancing substances. By the nineteenth century, endurance athletes like Abraham Wood used laudanum (an opiate) to stay awake during multiday competitions, and cyclists experimented with nitroglycerine to improve stamina.

3.1.2 Early Modern Era (1900s–1950s)

The 1904 Olympic marathon highlighted the risks of unregulated doping when runner Thomas Hicks nearly died after ingesting strychnine, brandy, and raw eggs. By the 1930s and 1940s, amphetamines became the stimulant of choice, and during World War II, anabolic steroids were tested on soldiers [2]. Post-war, these compounds were adapted for athletic and rehabilitation purposes. In the 1950s, Soviet athletes pioneered testosterone use, while U.S. physician John Ziegler developed Dianabol (methandrostenolone), marking the beginning of widespread synthetic steroid abuse.

3.1.3 *Cold War and State-Sponsored Programs (1960s–1980s)*

The death of Danish cyclist Knud Enemark Jensen at the 1960 Rome Olympics, linked to amphetamine use, prompted the International Olympic Committee to implement drug testing by 1968 [2]. East Germany's "State Plan 14.25" systematically administered steroids to athletes, particularly female swimmers, producing Olympic dominance but severe long-term health consequences. The 1976 Montreal Games revealed the ethical and physical costs of state-sponsored doping, with East German women capturing nearly all swimming golds.

3.1.4 *The Steroid Era and Scandals (1980s–1990s)*

Notable incidents, such as Canadian sprinter Ben Johnson's disqualification at the 1988 Seoul Olympics for stanozolol use and the 1998 Festina Affair at the Tour de France, underscored the pervasiveness of doping and spurred international reform. These events directly influenced WADA's formation and the establishment of unified anti-doping standards.

3.1.5 *Technological Arms Race and Modern Innovations (2000s–Present)*

The 2003 BALCO scandal, Lance Armstrong's disqualifications, and Russia's state-sponsored doping program exposed sophisticated evasion tactics, driving innovation in detection technologies.

WADA introduced the Athlete Biological Passport in 2004 to monitor biomarkers longitudinally, and carbon isotope ratio mass spectrometry (CIR-MS) was developed to differentiate synthetic from endogenous hormones [3]. Anti-doping efforts continued to evolve through the use of advanced instrumentation and specialized personnel, as demonstrated by the preparations for the 2024 Paris Olympics, which reflected a strong commitment to maintaining clean and fair competition.

3.2 International Bodies and Conventions Established to Combat Doping Across Sports

The global fight against doping in sport has evolved into one of the most coordinated and multidisciplinary efforts in modern international governance [6–10]. At its core, this system is built on a combination of international treaties, specialized organizations, and collaborative frameworks that work together to uphold fair play and safeguard the health of athletes. Over time, the landscape of anti-doping has shifted from fragmented national policies to a harmonized global regime in which governments, international sports bodies, and independent organizations collaborate closely.

The creation of the WADA in 1999 marked a turning point by introducing a central authority responsible for unifying rules across all sports and nations. In parallel, the International Testing Agency (ITA), launched in 2018, further strengthened operational independence in testing and results management. On the legal front, two landmark treaties the Council of Europe's Anti-Doping Convention (1989) and United Nations Educational, Scientific and Cultural Organization's (UNESCO) International Convention against Doping in Sport (2005) gave governments binding obligations to translate anti-doping rules into national law.

Together with the International Olympic Committee (IOC) Medical Commission, national agencies such as AFLD, specialized laboratories like the Laboratoire Antidopage Français (LADF), and regional cooperation mechanisms such as the Regional Anti-Doping Organizations (RADOs), this multi-layered system ensures both consistency and adaptability. Each entity has its own history, mandate, and mechanisms, yet they are united by a shared vision: to preserve the integrity of sport by eradicating doping practices.

3.2.1 Council of Europe Anti-doping Convention (Strasbourg, 1989)

The **Council of Europe Anti-Doping Convention** (ETS No. 135), opened for signature in Strasbourg on November 16, 1989, was the first legally binding international instrument aimed at eliminating doping in sport [11]. Long before the establishment of WADA or the adoption of the UNESCO Convention, this treaty laid the groundwork for coordinated state action by obliging signatory countries to adopt consistent laws and enforcement measures against doping.

Background and Adoption: The Convention emerged from concerns in the 1980s about increasing athlete health risks and the growing sophistication of performance-enhancing methods. Following a 1967 resolution by the Council of Europe's Committee of Ministers, the idea of a legal instrument gathered momentum. After several years of negotiations, the Convention was signed in both English and French (with both texts equally authoritative) and entered into force on March 1, 1990, once the required ten ratifications had been reached.

Key Provisions: The Convention contains several landmark provisions

- ***Scope:*** It applies to "doping in sport," defined broadly to include both the use of substances and methods that artificially enhance performance or endanger health.

- ***Common Terminology:*** For the first time, it introduced standardized definitions of categories such as anabolic agents, peptide hormones, and blood manipulation techniques.
- ***Legal Obligations:*** Parties must criminalize unauthorized manufacturing and trafficking of banned substances, empower national authorities to carry out inspections, and prosecute offenders.

Objectives: The Convention sought to eliminate doping through:

- ***Harmonization of National Laws***: Countries were required to establish legislation regulating substance control, prevention programs, testing mechanisms, and sanctions.
- ***Scientific Cooperation***: Member states were encouraged to share analytical methods, scientific research, and laboratory standards to ensure consistent testing practices.
- ***Enforcement Coordination***: The treaty facilitated mutual assistance, recognition of test results across borders, and transnational investigations.
- ***Monitoring:*** Implementation was overseen through periodic reports submitted to the Council of Europe's Steering Committee for Sport (CSP).

Signatures and Ratifications: When it was first introduced, 21 member states of the Council of Europe signed the treaty on its opening day. By 1990, the required number of signatories had been reached for it to take effect, and over time, participation expanded beyond Europe to include countries such as Australia, Canada, Morocco, and Tunisia. As of 2025, a total of 52 states have ratified or joined the treaty.

Additional Protocol (2002): An **Additional Protocol** (ETS No. 188) was adopted in 2002 to strengthen enforcement, particularly concerning the recognition of doping controls across borders. This Protocol required annual reports on measures taken, improved information sharing, and mutual recognition of testing results. It came into force in 2007 and remains a crucial supplement.

Implementation and Impact: The Convention influenced subsequent instruments, including the **World Anti-Doping Code (2003)** and the **UNESCO Convention (2005)**. Within Europe, it led to:

- Enhanced national legislation criminalizing doping activities.
- Accreditation of laboratories to internationally recognized standards (ISO/IEC 17,025).
- Peer-review mechanisms and evaluation visits by the CSP.

Even after more than three decades, the Convention remains a cornerstone of anti-doping governance, embodying the principle that states must not only cooperate with sports bodies but also legislate and enforce rules in their own jurisdictions.

3.2.2 World Anti-doping Agency (WADA)

The establishment of **WADA** in 1999 signaled the beginning of a truly global anti-doping framework [12]. Unlike earlier efforts that were confined to Europe or the Olympic Movement, WADA brought together governments and sporting organizations under a single authority.

History and Establishment: WADA was created in Lausanne on November 10, 1999, following the **Declaration of Lausanne,** which recognized the urgent need for harmonized anti-doping rules. Its headquarters moved to Montreal in 2002, while regional offices were established in Europe, Africa, Asia/Oceania, and Latin America to strengthen reach.

Mission and Governance: WADA's mission is "to lead a collaborative worldwide movement for doping-free sport." Its governance model is unique: the Foundation Board and Executive Committee are equally composed of representatives from the Olympic Movement and national governments. This structure ensures balanced decision-making and funding.

The World Anti-Doping Code: WADA's central instrument is the **World Anti-Doping Code,** which harmonizes rules across sports and countries. Updated every four years, the Code defines violations, sanctions, and testing standards. The latest revision entered into force in January 2021, with consultations already underway for the 2027 edition.

Key Functions and Programs

- ***Standards and Monitoring**:* WADA develops International Standards for testing, laboratory procedures, and therapeutic use exemptions.
- ***Laboratory Accreditation**:* Over 30 laboratories worldwide are accredited, each required to meet ISO/IEC 17,025 standards.
- ***Testing and Results Management**:* WADA oversees the Anti-Doping Administration & Management System **(ADAMS) platform** for secure data exchange and supports both targeted and intelligence-led testing.
- ***Education and Research**:* Through the Athlete Biological Passport Electronic Library **platform**, WADA delivers e-learning resources, while funding research into detection methods, emerging substances, and behavioral studies.
- ***Funding***: Its budget is split equally between the IOC and national governments, ensuring shared responsibility.

Impact and Future Directions: Since 1999, WADA has

- Unified more than 650 signatories under one Code.
- Advanced testing methods, including the Athlete Biological Passport.
- Expanded global education programs.

Looking forward, WADA aims to adopt AI-driven detection methods, integrate next-generation mass spectrometry tools, and strengthen compliance monitoring.

3.2.3 International Testing Agency (ITA)

The ITA, created in 2018, is a non-profit body that provides independent testing services to international federations and event organizers [13].

Establishment and Governance: The ITA was founded as a Swiss-law foundation after the 2015 Olympic Summit recommended insulating anti-doping operations from conflicts of interest. It is overseen by an independent Board of Directors and operates separately from the IOC and WADA, though funded by both.

Core Functions:

- *Testing:* Conducts both urine and blood sample collection in-competition and out of competition.
- *Results Management:* Reviews adverse findings and issues sanctions under the World Anti-Doping Code.
- *Athlete Whereabouts:* Manages registered testing pool athletes through ADAMS.
- *Training:* Runs international certification programs for Doping Control Officers.
- *Laboratory Oversight:* Partners with WADA-accredited labs to ensure compliance with global standards.

Impact and Future Work: Since 2018, the ITA has overseen over 200,000 tests, managed anti-doping programs for Olympic Games, and uncovered numerous violations through re-analysis of stored samples. Future priorities include intelligence-driven testing and expanding collaboration with law enforcement.

3.2.4 French Anti-doping Agency (AFLD)

The Agence française de lutte contre le dopage (AFLD) is France's independent authority responsible for regulating and enforcing anti-doping measures in sport [14]. Established by the 2006 "Lamour Law," the agency operates autonomously from government bodies and sports federations, ensuring neutrality in its mission. As the country's central anti-doping institution, the AFLD aligns its operations with the World Anti-Doping Code and works to promote clean and fair competition across all levels of sport.

Core Mandates and Activities: The AFLD conducts testing both in and out of competitions, applying a risk-based strategy that accounts for the demands of each sport, competition schedules, and historical data on doping. All collected samples are analyzed by the French Anti-Doping Laboratory (Laboratoire Antidopage Français, LADF) at the University of Paris-Saclay, which holds WADA accreditation. In 2023 alone, more than 12,000 samples were examined at this facility.

Since 2019, the AFLD has held exclusive authority to adjudicate doping violations in France. Suspected breaches are first reviewed by the agency's College, which may recommend sanctions. If contested, cases are passed to an independent Sanctions Committee for final judgment.

Investigations and Intelligence Gathering: The agency's mandate extends beyond testing. AFLD pursues non-analytical violations such as trafficking, complicity, or attempted doping. Its investigators work closely with police, customs, and judicial authorities, with legal authority to summon individuals, inspect professional premises, and, with court approval, search private residences. The agency also employs digital surveillance to identify traffickers and suppliers.

Education and Athlete Engagement: Prevention and education form another key pillar of AFLD's strategy. Since 2018, its education department has designed programs for athletes, coaches, and medical staff, while training "ambassadors" to spread anti-doping awareness within sports communities. In 2019, AFLD established a Sports Committee made up of elite athletes, who provide input on policies, participate in outreach, and ensure that athletes' perspectives are represented nationally and internationally.

International Cooperation and Scientific Capacity: AFLD actively partners with global stakeholders, including WADA and the International Testing Agency (ITA). Ahead of the Paris 2024 Olympic and Paralympic Games, it signed a Memorandum of Understanding with ITA to coordinate testing and intelligence sharing with French authorities. At the same time, LADF expanded its capacity to process around 8,000 Olympic samples, with a workforce of 100 specialists. Known for pioneering methods to detect EPO and insulin analogues, LADF represents one of Europe's leading anti-doping laboratories.

Legal and Extended Jurisdiction: AFLD operates under the French Sports Code and benefits from legislative milestones, from the 1999 "Buffet Law" to the 2021 ordinance strengthening its investigative and educational powers. Its mandate also covers animal doping in equestrian and canine events, reflecting the broad scope of its responsibility in safeguarding fairness across all forms of sport.

3.2.5 UNESCO International Convention Against Doping in Sport (Paris, 2005)

Adopted in Paris on October 19, 2005, UNESCO's International Convention against Doping in Sport is the first legally binding global treaty that commits governments to implement anti-doping measures consistent with the World Anti-Doping Code [15]. It officially entered into force on February 1, 2007 following its 30th ratification, and by October 2022 had been ratified by 189 Parties, embedding anti-doping obligations into national law on an unprecedented scale.

Background and Purpose: The Convention was unanimously endorsed at UNESCO's 33rd General Conference, reflecting a shared recognition that doping undermines athlete health, damages fair competition, and erodes the integrity of sport. Its primary objective is to promote harmonized legislation, coordinated testing, and international collaboration to eliminate doping practices. Unlike voluntary codes of conduct, this treaty requires governments to integrate its provisions into domestic law, thereby aligning legal and sporting frameworks worldwide.

Relationship with the World Anti-Doping Code: While the Convention is grounded in the principles of the World Anti-Doping Code, it goes further by making compliance a matter of international law. States must incorporate the Code's standards into their legal systems but retain the flexibility to introduce complementary measures. Its annexes, unlike the Code's appendices, carry legally binding force, strengthening the global enforcement framework.

Key Obligations of Member States: Parties to the Convention are required to:

- Legislate and regulate against the production, trafficking, possession, and administration of banned substances and methods.
- Support anti-doping organizations such as WADA, international federations, and national bodies in conducting effective in- and out-of-competition testing.
- Cooperate internationally, recognizing each other's test results, sharing intelligence, and assisting with cross-border enforcement.
- Invest in education and research, raising awareness of health risks and ethics while promoting innovation in detection methods.
- Ensure compliance through funding mechanisms, including withholding public resources from non-compliant sports organizations and contributing equally with the Olympic Movement to WADA's budget.

Governance and Monitoring: The treaty is overseen by the Conference of Parties (COP), which meets every 2 years to assess progress, review implementation reports, and adopt updates. Recent decisions include the development of optional protocols to address new threats such as genetic doping.

Impact and Legacy: The Convention has created near-universal alignment of national anti-doping laws with international standards. By formalizing government responsibilities, it extends the reach of WADA and ensures that anti-doping measures are recognized not only within sport but also within judicial and regulatory systems. Its continuing evolution, particularly through new protocols, ensures that global frameworks remain responsive to emerging scientific and ethical challenges.

3.2.6 International Olympic Committee (IOC) Medical Commission

The IOC Medical Commission was created to safeguard athlete health, guide anti-doping policies, and provide a strong scientific foundation for sport [16]. Established in 1967 under the leadership of Alexandre de Merode, the Commission quickly became the central body uniting the Olympic Movement's medical and anti-doping activities. Over the decades, it has been responsible for setting medical standards, implementing doping controls, drafting the IOC Medical Code, and advancing sports medicine research through collaborations and publications. Its contributions range from delivering daily medical oversight at the Games to driving long-term initiatives that shaped modern anti-doping systems and athlete welfare protocols.

Historical Development: The roots of the Commission trace back to the early 1960s, when IOC President Avery Brundage highlighted the growing dangers of doping. An initial anti-doping body was created in 1962 under Arthur Porritt, but it remained inactive until 1967. At the 66th IOC Session in Tehran, Alexandre de Merode was appointed Chair, formally establishing the IOC Medical Commission. Under his leadership (1967–2002), the Commission expanded its structure after the Montreal 1976 Games, forming sub-commissions to focus on doping and biochemistry, biomechanics, physiology, orthopaedics, and cooperation with National Olympic Committees (NOCs).

Mandate and Governance: The Commission's primary role is advisory, providing expertise to the IOC Session, Executive Board, and President on athlete health, anti-doping efforts, medical services at the Games, and research priorities. It is also responsible for drafting and updating the IOC Medical Code, which defines prohibited substances and methods, athlete obligations, and sanction procedures. Its governance structure includes leading experts in sports medicine, with members nominated by NOCs, International Federations, and professional medical bodies.

Core Activities: A major responsibility of the Commission is anti-doping oversight. It works with organizing committees and federations to design and supervise doping-control programs both in and out of competition. Its sub-commission on doping and biochemistry established accreditation standards for laboratories, ensuring consistent quality worldwide. Beyond anti-doping, the Commission organizes medical services during the Games, including emergency response, daily athlete care, and coordination with local health providers. It also oversaw gender verification procedures for female athletes, issuing certificates recognized across IOC events.

On the research front, the Commission has played a leading role in publishing the Encyclopaedia of Sports Medicine and in supporting studies on injury prevention, rehabilitation, and performance science through collaborations such as the Association of Olympic International Research in Sports Medicine (AOIRMS).

Impact and Legacy: The IOC Medical Commission laid the groundwork for today's global anti-doping system, influencing WADA and the UNESCO Convention. It also set enduring standards for athlete medical care, ensuring consistent protocols across events. Through its research and publications, it has advanced evidence-based sports medicine worldwide.

3.2.7 Regional Anti-doping Organizations (RADOs)

Regional Anti-Doping Organizations (RADOs) were created under the framework of the WADA to strengthen global capacity in the fight against doping and ensure fair competition [17]. Launched in 2004, the RADO Program was designed to support regions where National Anti-Doping Organizations (NADOs) were underdeveloped or lacked sufficient resources. By providing expertise in testing, results management, education, and policy development, RADOs have become essential in bridging gaps between countries while ensuring consistent application of the World Anti-Doping Code.

RADO Program and Structure: The RADO Program operates with the overarching goal of protecting clean sport through regionally tailored anti-doping initiatives. WADA provides strategic oversight, technical guidance, and funding, while governance is handled locally by participating countries. Each RADO is typically managed by a board made up of representatives from member states, formalized through agreements with governments, National Olympic Committees, and regional sports federations. This hybrid model allows RADOs to respect cultural and regional contexts while still aligning with international standards.

Core Services: RADOs deliver a wide range of services designed to strengthen anti doping systems across multiple countries:

- *Testing and Results Management:* Coordinating both in-competition and out-of-competition testing, sharing resources to reduce costs while ensuring reliable results.
- *Education and Outreach*: Designing athlete- and coach-focused programs aligned with WADA's International Standard for Education, often adapted into local languages and cultural frameworks.
- *Policy and Research Support:* Conducting region-specific studies on doping trends, which feed into broader WADA policy development.
- *Therapeutic Use Exemptions (TUEs):* Running regional panels to review medical exemptions consistently across multiple nations.
- *Capacity Building:* Training and certifying Doping Control Officers (DCOs) and other specialists, often in collaboration with WADA's online learning platform (ADEL).

Regional Examples

- *Africa Zone VI RADO:* Established in 2006 in Mozambique, it coordinates anti-doping activities across 10 Southern African countries, providing leadership in testing, TUE management, and education.
- *Caribbean RADO:* Based in Barbados, it serves 18 island nations, tailoring education programs and centralizing testing operations to meet the unique needs of the region.

- *Pan-American RADO (PAN-RADO):* Supports Central and South America by offering services in Spanish and Portuguese, ensuring wider accessibility and compliance.

Impact and Future Outlook: RADOs have played a crucial role in expanding testing coverage, promoting education, and elevating anti-doping awareness in regions previously underserved. Many countries have advanced to establishing independent NADOs, reducing reliance on regional bodies. Looking ahead, RADOs are expected to embrace intelligence-led testing, digital monitoring, and remote education, ensuring that the global fight against doping remains adaptive, inclusive, and effective.

3.3 International Organizations and Committees Combating Doping in Animal Sports

The regulation of doping in animal sports has become a vital issue at the intersection of fair play, scientific progress, and animal welfare. While human anti-doping policies are well established under frameworks such as the World Anti-Doping Code, similar vigilance is required in animal-centered competitions. Sports such as horse racing, equestrian disciplines, and greyhound racing rely heavily on the physical performance of animals, making them vulnerable to the misuse of substances that can artificially enhance stamina, speed, or recovery. This not only jeopardizes fair competition but also poses serious risks to animal health and safety [18–23].

To address these challenges, a network of international organizations has been created to standardize testing, enforce regulations, and promote education. These organizations, including the International Federation of Horseracing Authorities (IFHA), the Fédération Équestre Internationale (FEI), and several regional and national anti-doping bodies, collaborate closely to align scientific, veterinary, and regulatory practices. Their collective mission is to foster integrity in animal sports while safeguarding the welfare of animals who are at the heart of these competitions.

3.3.1 International Horseracing Authorities (IFHA)

The IFHA is recognized as the global governing body for Thoroughbred horse racing [24]. Established in 1993 and headquartered in Paris, France, IFHA coordinates the activities of more than 60 national racing authorities and regional federations, such as the Asian Racing Federation. Its mission is anchored in three core objectives: ensuring fairness and integrity in competition, promoting the global development of the sport, and safeguarding the welfare of horses and jockeys.

Mission and Objectives: The IFHA works to promote international cooperation in racing while harmonizing regulatory standards. Its goals include:

- Fostering the worldwide growth of Thoroughbred racing.
- Upholding equine and human welfare standards.
- Enforcing integrity through anti-doping regulations.
- Standardizing race classifications and rankings.
- Collaborating with international partners such as the Fédération Équestre Internationale (FEI) and the World Organisation for Animal Health (WOAH).

International Agreement on Breeding, Racing, and Wagering (IABRW): The IABRW forms the backbone of IFHA's governance structure. It outlines

- Lists of prohibited substances and methods.
- Testing protocols, both during and outside competition.
- Standards for race classification (e.g., Group and Grade races).
- Breeding and stud book regulations.
- Integrity measures for wagering and betting.

Article 6 of the IABRW explicitly deals with anti-doping, reaffirming the principle of zero tolerance for prohibited substances.

Anti-Doping and Laboratory Accreditation: The IFHA maintains an extensive network of accredited laboratories, located in countries such as Australia, France, Great Britain, Hong Kong, Japan, and the United States. These laboratories operate under stringent performance standards to ensure reliable and reproducible results. The IFHA collaborates with the International Laboratory Accreditation Cooperation (ILAC) to harmonize analytical methods, which is crucial for maintaining global consistency.

Welfare Commitments and Global Rankings: Beyond anti-doping, IFHA emphasizes equine welfare. Its Horse Welfare Committee has published Minimum Horse Welfare Standards, offering guidance on care from foaling to retirement. Additionally, IFHA manages internationally recognized rankings, including the Longines World's Best Racehorse Rankings and the Longines World's Best Jockey Rankings, which standardize the evaluation of excellence in the sport.

Governance and Structure: The IFHA's governance is supported by specialized committees

- Advisory Council on Equine Prohibited Substances and Practices.
- International Grading and Race Planning Advisory Committee.
- Committee for the Harmonization of Racing Rules.
- Horse Welfare Committee.

Through these structures, IFHA shapes policy, ensures regulatory uniformity, and responds to emerging challenges in international horse racing.

3.3.2 *International Federation for Equestrian Sports (FEI)*

The International Federation for Equestrian Sports (FEI), or Fédération Équestre Internationale, is the world governing body for equestrian disciplines [25]. Founded in 1921 in Lausanne, Switzerland, the FEI supervises Olympic and non-Olympic disciplines, establishes regulatory frameworks, and promotes the welfare of both horses and riders. Today, the organization brings together 136 national federations under a shared mission of fostering equestrian sport in a fair, ethical, and sustainable way.

History and Evolution: The FEI was created following the inclusion of equestrian events Jumping, Dressage, and Eventing in the 1912 Olympic Games in Stockholm. Eight countries initially formed the FEI to establish standardized rules. Over time, membership grew, and the FEI expanded its remit to cover new disciplines, including Para-Equestrian sports.

Disciplines Governed: The FEI regulates both Olympic and non-Olympic disciplines

- *Olympic*: Jumping, Dressage, Eventing.
- *Non-Olympic*: Driving, Endurance, Vaulting.
- *Para-Equestrian*: Para Dressage and Para Driving.

The FEI works in collaboration with partner organizations for disciplines such as Horseball and Tent Pegging; however, it does not govern horse racing or polo.

Anti-Doping Framework: A cornerstone of the FEI's mission is its Equine Anti-Doping and Controlled Medication Programme (EADCMP). First codified in 2004 and currently in its fourth edition (effective 2025), the program defines prohibited substances, establishes sample collection protocols, and maintains laboratory accreditation standards. The FEI Prohibited Substances List is updated annually and aligns closely with WADA's list for human athletes, ensuring parity across the broader sporting world.

Testing and Enforcement: The FEI uses a wide range of testing methods including blood, urine, hair, and saliva samples to detect banned substances. It categorizes substances into two groups:

- *Banned Substances:* Prohibited at all times, with no therapeutic justification.
- *Controlled Medications:* Prohibited during competition but may be permitted for legitimate veterinary purposes outside competition.

Sanctions for violations range from fines and disqualifications to suspensions, depending on the severity of the offense.

Governance and Integrity Standards: The FEI's governance model emphasizes transparency and ethics. Its Veterinary Department manages the EADCMP, while its disciplinary structures adjudicate cases of doping and misconduct. The FEI has also introduced measures such as a Social Media Policy and integrity guidelines to safeguard the sport's reputation.

Laboratory Standards and Accreditation: FEI-approved laboratories operate under harmonized testing protocols, ensuring consistency worldwide. Directors of these laboratories form the FEI Laboratory Group, which collaborates with the Veterinary Department to enhance testing technologies and maintain reliability in anti-doping analysis.

Horse Welfare and Ethical Standards: The FEI upholds the Code of Conduct for the Welfare of the Horse, a set of principles ensuring horses are treated with dignity and respect both in and out of competition. Welfare considerations are integrated into every regulatory update, reinforcing the organization's commitment to the humane treatment of equine athletes.

Global Impact: With its headquarters in Lausanne and a worldwide network of national federations, the FEI wields considerable influence over the future of equestrian sports. Its frameworks for anti-doping, laboratory accreditation, and horse welfare serve as benchmarks for other animal sports. Through its blend of regulatory authority, scientific expertise, and welfare advocacy, the FEI continues to shape equestrian disciplines on a global scale.

Collaborative Networks in Animal Sports: Although IFHA and FEI serve as primary international regulators in horse-related sports, their efforts are complemented by national agencies such as the French Anti-Doping Agency (AFLD), industry certification bodies like the Banned Substances Control Group (BSCG), and regional authorities. Together, these organizations create a multi-layered system of oversight that emphasizes scientific rigor, fair competition, and ethical responsibility.

Their collaborations involve:

- Standardizing testing methods across laboratories.
- Sharing intelligence on emerging doping threats.
- Coordinating education initiatives for veterinarians, trainers, and owners.
- Establishing global databases of testing results and violations.

Through this web of partnerships, the global sporting community ensures that animal sports remain credible, safe, and respectful of animal welfare.

3.3.3 *Association of Racing Commissioners International (ARCI)*

The ARCI, founded in 1934, is one of the longest standing and most influential regulatory organizations in the field of horse and greyhound racing [26]. Over the decades, ARCI has evolved into the principal coordinating body for regulatory agencies, working to safeguard integrity, fairness, and transparency across a highly competitive industry that spans multiple jurisdictions.

Mission and Governance Structure: At the heart of ARCI's mission is the protection of both the public interest and the participants of the racing industry, including owners, trainers, jockeys, bettors, and the animals themselves. Recognizing that racing is a sport with international reach but strong local variations, ARCI strives to harmonize standards so that rules are not only effective but also consistently applied across states, provinces, and even countries.

The organization supports its member agencies through:

- *Policy Formation:* ARCI develops model policies and provides evidence-based guidance to help regulatory authorities update or refine their local frameworks.
- *Information Sharing:* It facilitates a platform where members exchange regulatory data, racing results, scientific updates, and enforcement strategies.
- *Education and Training:* ARCI organizes workshops, training modules, and professional resources to build capacity among regulators and stewards.
- *Integrity Advocacy:* It promotes reforms designed to reinforce public trust and to maintain racing as a fair and credible sport.

Membership is drawn from dues-paying racing commissions and their staff, giving them direct access to ARCI services and decision-making. In addition, industry stakeholders such as veterinary groups, breeders' associations, and technology providers can participate as Associate Members, broadening the scope of collaboration.

Model Rules of Racing: One of ARCI's greatest contributions is the Model Rules of Racing, which serve as a standardized framework for jurisdictions around the world. These rules act as a regulatory "blueprint," ensuring that the conduct of racing, medication use, wagering procedures, and welfare practices are consistent across state or national borders.

The Model Rules cover:

- *Medication and Testing:* Clear protocols for therapeutic substances, withdrawal times, and prohibited drugs.
- *Operational Guidelines:* Procedures for starting races, adjudicating disputes, handling wagers, and disciplining offenders.
- *Safety Provisions:* Rules designed to safeguard jockeys, trainers, and horses from avoidable risks.

These rules are updated regularly to reflect advances in veterinary science and sports regulation. For example, the 2024 release of Version 13.1 introduced new standards for safety equipment and enhanced track protocols, underscoring ARCI's responsiveness to emerging challenges.

Medication and Drug Testing Standards: A defining feature of ARCI's work is its collaboration with the Racing Medication and Testing Consortium (RMTC), which focuses on establishing medication control standards. Together, ARCI and RMTC provide:

- A Controlled Therapeutic Medication Schedule, outlining what can be administered and under what conditions.

- Uniform Classification Guidelines, ranking substances by their effect on performance and recommending penalties based on severity.
- Laboratory Standards, requiring accredited labs to meet stringent criteria for testing reliability and transparency.

By setting uniform rules, ARCI helps to create a level playing field across jurisdictions, while also prioritizing the health and welfare of the racehorses involved.

International Collaboration and Expansion into Betting: Although ARCI's primary focus lies in North American racing, the organization maintains strong ties with global counterparts such as IFHA and FEI. Through these partnerships, ARCI contributes to worldwide discussions on doping controls, welfare policies, and regulatory convergence.

ARCI also demonstrates flexibility in adapting to the changing landscape of sports regulation. With the legalization of sports betting in many U.S. states, ARCI established the Sports Betting Regulators Association (SBRA), a subsidiary launched in 2024. The SBRA helps state agencies develop regulatory systems for sports wagering, highlighting ARCI's expanding influence beyond traditional racing.

3.3.4 French Anti-doping Agency (AFLD)—Animal Sport

The French Anti-Doping Agency (Agence Française de Lutte contre le Dopage—AFLD) was founded in 2006 as an independent authority overseeing compliance with anti-doping rules in France [27]. While the AFLD is widely recognized for its role in human athletics, its mandate also extends into animal sports, particularly equestrian events, sled dog racing, and canine competitions.

Jurisdiction and Scope: Under Article L. 241-1 of the French Sports Code, the AFLD is responsible for implementing veterinary anti-doping measures. Its jurisdiction, however, excludes horse racing, which is separately regulated by the Fédération Nationale des Courses Hippiques and the Ministry of the Interior. Instead, the AFLD works with organizations such as:

- Fédération Française d'Équitation (FFE).
- Fédération des Sports et Loisirs Canins.
- Fédération Française des Sports de Traîneau.

Through these partnerships, the AFLD ensures that doping control extends beyond traditional human sport to cover diverse animal competitions.

Veterinary Oversight and Sample Collection: One of AFLD's distinguishing features is its reliance on specially trained veterinarians. These professionals collect samples both in competition and out of competition, ensuring comprehensive monitoring. In 2023, the agency expanded its veterinary team by accrediting nine new practitioners, increasing its capacity to conduct surprise checks and meet growing demands.

The AFLD's work goes beyond regulation; it also provides support services to sporting organizers. For instance, collaborations with the Société Hippique Française (SHF) and the Conseil International du Sport Militaire (CISM) illustrate how the agency integrates seamlessly into the broader sporting ecosystem.

Legal Framework and Investigative Powers: French law provides the AFLD with robust legal authority, particularly after the November 2021 law against animal abuse. Investigators can:

- Access stables, kennels, and training facilities.
- Collect explanations or testimonies directly from stakeholders.
- Conduct undercover operations, including controlled "test purchases" of prohibited substances.

This framework equips the AFLD with tools to combat not just doping itself, but also the trafficking and distribution networks that support it.

Unique Features of Animal Anti-Doping: Although the AFLD borrows principles from human anti-doping systems, its animal program has unique features:

- Only owners, trainers, and athletes can be sanctioned; veterinarians and caretakers are not directly liable.
- There is no global standard for sanction duration, and penalties tend to be less severe than in human sports.
- Unlike human sports, there are no formal therapeutic use exemptions for veterinary treatments.
- The statute of limitations for disciplinary proceedings is 5 years, compared to ten years in human doping cases.

Such differences reflect the complexities of adapting a human-centered regulatory system to the realities of animal sport.

Collaborative Role in European Anti-Doping: The AFLD cooperates with law enforcement and judicial authorities, particularly during major events such as the 2024 Paris Olympics, where animal sports drew international attention. Although no single global anti-doping framework exists for animals, France's AFLD has emerged as a pioneering model, balancing rigorous enforcement with an adaptable regulatory approach.

3.3.5 World Anti-doping Agency (WADA)—Animal Sport Collaboration

While the WADA was established to regulate human sports, its influence indirectly extends into the animal domain [12]. This is particularly evident in equestrian disciplines governed by the FEI, which has voluntarily adopted WADA's standards and methodologies.

Integration of WADA Standards: The FEI's Equine Anti-Doping and Controlled Medication Programme (EADCMP) mirrors many of WADA's frameworks. It includes

- A Prohibited Substances List, aligned with WADA's guidance.
- Standardized testing and collection protocols, including blood, urine, and hair samples.
- Comprehensive educational resources, ensuring trainers, riders, and veterinarians understand compliance requirements.

Through this alignment, equestrian sports benefit from the credibility and global recognition associated with WADA's work in human athletics.

Addressing Emerging Threats: One pressing concern in both human and animal doping is gene doping. WADA, in partnership with veterinary authorities, is at the forefront of developing detection technologies and databases that track genetic manipulation. The emphasis remains not only on detection but also on ensuring that procedures respect the welfare of animals.

Broader Collaborations: WADA's partnerships with organizations like the World Health Organization (WHO) and the International Federation of Pharmaceutical Manufacturers & Associations (IFPMA) primarily focus on human sports, yet the knowledge transfer directly benefits animal doping control. For example, early identification of substances in pharmaceutical development accelerates the creation of new equine and canine testing methods.

Thus, while WADA does not directly regulate animal doping, its frameworks act as a global reference point, ensuring that animal sports remain aligned with broader anti-doping efforts.

3.3.6 Greyhound Board of Great Britain (GBGB)

The GBGB is the governing body responsible for overseeing licensed greyhound racing in the UK [28]. Created in 2009 through the merger of the British Greyhound Racing Board and the National Greyhound Racing Club, the GBGB combines regulatory oversight with a strong commitment to animal welfare.

Governance and Integrity Oversight: GBGB's regulatory functions are anchored in its Integrity and Compliance Department, which manages doping control, race licensing, and disciplinary procedures. Its Doping and Medication Review Panel (DMRP), active since 2012, sets the strategic direction for anti-doping policies and recommends updates based on advances in veterinary pharmacology.

The organization maintains UKAS accreditation, ensuring that testing procedures meet high international standards. Oversight is provided by a diverse board representing track promoters, trainers, veterinary experts, and independent directors, balancing industry perspectives with regulatory independence.

Welfare Strategy: **"A Good Life for Every Greyhound"**: GBGB's landmark welfare initiative, launched in 2022, sets ambitious goals for the lifetime care of racing greyhounds. Its principles cover

- *Veterinary Oversight:* Independent veterinarians must be present at all licensed tracks.
- *Track Safety:* Regular inspections ensure surfaces meet safety standards and that racing is suspended in extreme weather.
- *Kennel Conditions*: Trainers must comply with PAS 251 standards, with independent audits verifying conditions.
- *Retirement and Rehoming:* Through the Greyhound Retirement Scheme, 94% of retired greyhounds are successfully rehomed, eliminating unnecessary euthanasia.

Funding and Calls for Statutory Levy: Currently, welfare initiatives rely heavily on voluntary contributions from bookmakers, set at 0.6% of off-course betting turnover. However, the voluntary nature of this arrangement creates uncertainty. Consequently, GBGB has campaigned for a statutory levy, ensuring a more sustainable and predictable funding model for greyhound welfare.

Challenges and Future Outlook: Although GBGB has achieved progress in welfare and integrity, the sport faces challenges such as declining popularity and revenue pressures. Nevertheless, GBGB's continued commitment to transparency, modernization, and welfare-first approaches positions it as a central player in shaping the sport's future.

3.3.7 Greyhound Racing Ireland (GRI)

GRI, formerly known as the Irish Greyhound Board (IGB), regulates and promotes greyhound racing within the Republic of Ireland [29]. Established under the Greyhound Industry Act of 1958, the organization was rebranded in 2020 as Rásaíocht Con Éireann (RCÉ), reflecting its modernization agenda and commitment to animal welfare.

Structure and Operations: GRI manages 17 licensed tracks, nine of which are state operated, while the rest are privately managed. The organization is responsible for

- Licensing trainers, officials, and bookmakers.
- Operating Tote betting systems and applying levies.
- Funding prize money, facility upgrades, and marketing.

Commitment to Welfare: GRI has responded to heightened public scrutiny with several reforms:

- *Greyhound Care Fund (2019):* A welfare-focused fund supporting injury treatment, rehoming, and rehabilitation.

- *Traceability System (2021):* Tracks greyhounds from birth to retirement, increasing accountability.
- *Kennel Improvement Scheme (2022–2024):* Provides grants to improve living standards in kennels.
- *Expanded Welfare Inspections:* Over 2,600 inspections were conducted in 2022 alone.

Addressing Public Concerns: Following critical media reports in 2019, particularly on overbreeding and euthanasia, GRI committed to transparency and reform. While challenges remain, the organization emphasizes that malpractice is the work of a minority and continues to strengthen safeguards for animal welfare.

3.4 Summary

This chapter has provided a comprehensive exploration of doping in sport, tracing its historical evolution, regulatory frameworks, and global governance mechanisms. Beginning with the origins of performance enhancement and the ethical dilemmas it has posed, the discussion highlighted the pivotal milestones that shaped modern anti-doping practices, including the creation of WADA, the adoption of the World Anti-Doping Code, and the implementation of UNESCO's International Convention against Doping in Sport. The chapter also examined the roles of national and regional anti-doping organizations (NADOs and RADOs), illustrating how coordinated testing, education, and intelligence-led investigations ensure consistent enforcement and promote athlete welfare.

Special attention was given to doping in animal sports, encompassing horse racing, equestrian competitions, and greyhound racing. The chapter outlined the contributions of international organizations such as the IFHA, FEI, GBGB, AFLD, and GRI, emphasizing their mandates, laboratory standards, and anti-doping initiatives that safeguard both competitive integrity and animal welfare. Case studies and examples demonstrated the importance of collaboration, harmonized regulations, and technological advancement in detecting prohibited substances and adapting to emerging challenges such as gene doping.

In conclusion, this chapter underscores that combating doping is a multidimensional effort integrating science, ethics, law, and education. By providing a detailed historical context, analyzing regulatory frameworks, and highlighting global governance strategies, the book establishes a holistic understanding of the fight against doping. It reinforces that protecting fairness, health, and integrity in sport human or animal is a shared global responsibility, requiring ongoing vigilance, research, and international cooperation.

References

1. Vlad RA, Hancu G, Popescu GC, et al. Doping in sports, a never-ending story? Adv Pharm Bull. 2018;8:529–534. https://doi.org/10.15171/apb.2018.062.
2. Holt RI, Erotokritou-Mulligan I, Sönksen PH. The history of doping and growth hormone abuse in sport. Growth Horm IGF Res. 2009;19:320–6. https://doi.org/10.1016/j.ghir.2009.04.009.
3. World Anti-Doping Code. https://www.wada-ama.org/sites/default/files/resources/files/2021_wada_code.pdf. Accessed 6 May 2025.
4. United States' and International Response to the Problem of Doping in Sports. https://ir.lawnet.fordham.edu/cgi/viewcontent.cgi?article=1244&context=iplj. Accessed 6 May 2025.
5. Holz M, Robertson J. How to develop intelligence gathering in efficient and practical anti-doping activities. Med Sport Sci. 2017;62:139–52. https://doi.org/10.1159/000460724.
6. Ljungqvist A. Brief history of anti-doping. Med Sport Sci. 2017;62:1–10. https://doi.org/10.1159/000460680.
7. Young R. The development of the world anti-doping code. Med Sport Sci. 2017;62:11–21. https://doi.org/10.1159/000460681.
8. Bowers LD. Science and the rules governing anti-doping violations. Handb Exp Pharmacol. 2010;195:513–31. https://doi.org/10.1007/978-3-540-79088-4_21.
9. Ivanova V, Boghosian T, Rabin O. The WADA proficiency testing program as an integral part of the fight against doping in sport. Accred Qual Assur. 2007;12:491–3. https://doi.org/10.1007/s00769-007-0283-4.
10. Meija J, Possolo A, Garrido BC, et al. New statistical framework for interlaboratory evaluation of anti-doping testing results by WADA. Accred Qual Assur. 2024;29:345–57. https://doi.org/10.1007/s00769-024-01595-w.
11. Council of Europe Anti-Doping Convention. https://www.coe.int/en/web/sport/anti-doping-convention. Accessed 6 May 2025.
12. World Anti-Doping Agency. https://www.wada-ama.org/en. Accessed 6 May 2025.
13. International Testing Agency (ITA). https://ita.sport/. Accessed 6 May 2025.
14. French Anti-Doping Agency (AFLD). https://www.afld.fr/en/. Accessed 6 May 2025.
15. UNESCO International Convention against Doping in Sport (Paris, 2005). https://www.unesco.org/en/legal-affairs/international-convention-against-doping-sport. Accessed 7 May 2025.
16. International Olympic Committee (IOC) Medical Commission. https://www.olympics.com/ioc/health-medicine-and-science-commission. Accessed 7 May 2025.
17. Regional Anti-Doping Organizations (RADOs). https://www.wada-ama.org/en/anti-doping-partners/regional-anti-doping-organizations-rados. Accessed 7 May 2025.
18. Animal Wellness Action. Protecting America's horses (2024). https://animalwellnessaction.org/protecting-americas-horses/. Accessed 8 May 2025.
19. Humane Society of the United States. Humane society urges horse racing industry to embrace protective measures (2024). https://humaneaction.org/press-release/2024/04/humane-society-united-states-urges-horse-racing-industry-embrace-protective. Accessed 8 May 2025.
20. International Federation of Sleddog Sports (IFSS). Doping control. https://sleddogsport.net/eng/81. Accessed 8 May 2025.
21. Horseracing Integrity and Safety Authority (HISA). Anti-doping program goes into effect for horse racing (2023). https://www.avma.org/news/anti-doping-program-goes-effect-horse-racing. Accessed 12 May 2025.
22. Greyhound Coalition. Doping and drugging in greyhound racing. https://greyhoundcoalition.com/media-resources/doping-and-drugging/. Accessed 12 May 2025.
23. American Society for the Prevention of Cruelty to Animals (ASPCA). Greyhound racing: animal cruelty and take action. https://www.aspca.org/improving-laws-animals/public-policy/greyhound-racing. Accessed 12 May 2025.
24. International Horseracing Authorities (IFHA). https://www.ifhaonline.org/. Accessed 12 May 2025.
25. International Federation for Equestrian Sports (FEI). https://www.fei.org/. Accessed 12 May 2025.

26. Association of Racing Commissioners International (ARCI). https://www.arci.com/. Accessed 12 May 2025.
27. French Anti-Doping Agency (AFLD)—Animal Sport. https://www.afld.fr/en/the-fight-against-doping/animal-doping/. Accessed 12 May 2025.
28. Greyhound Board of Great Britain (GBGB). Rules of racing. https://www.gbgb.org.uk/rules-regulation/rules-of-racing/. Accessed 12 May 2025.
29. Greyhound Racing Ireland (GRI). https://www.gov.ie/en/greyhound-racing-ireland/. Accessed 12 May 2025.

Chapter 4
The Prohibited Framework: Ensuring Safety and Fair Competition

Abstract In this chapter, the discussion centers on the critical framework of prohibited substances and methods within global sports governance, highlighting their significance in safeguarding athlete health, maintaining fair play, and preserving the integrity of competition. The historical trajectory of anti-doping regulation is traced from early Olympic concerns in the 1960s to the establishment of the WADA in 1999 and the harmonization of international rules under the 2004 World Anti-Doping Code. At the heart of this framework lies the annually updated Prohibited List, which categorizes substances and methods into distinct sections based on their physiological impact, performance-enhancing potential, health risks, and ethical implications. This chapter explains how different classes of substances, such as anabolic agents, peptide hormones, stimulants, narcotics, and masking agents, can artificially enhance performance while exposing athletes to serious medical risks ranging from cardiovascular disorders to psychological complications. Beyond chemicals, prohibited methods including blood manipulation, chemical tampering, and gene or cell doping are examined as evolving threats to fairness and health. The discussion also highlights the complexities introduced by therapeutic use exemptions (TUEs) and the ongoing struggle to detect designer drugs crafted to evade testing. By analyzing the rationale, mechanisms, and enforcement strategies surrounding doping, this chapter underscores the broader ethical dimension: the need to protect the spirit of sport as a celebration of natural talent, discipline, and human potential. As science advances, anti-doping regulation must remain vigilant, adaptive, and globally coordinated. Ultimately, this chapter frames prohibition not as restriction alone, but as a collective safeguard ensuring that competitive sports remain an authentic and inspiring human endeavor.

Keywords Prohibited substances · Anti-doping regulations · World anti-doping agency · WADA · Performance enhancement · Athlete health and safety · Therapeutic use exemptions · Prohibited methods · Fair competition in sport · Gene and cell doping · Sports integrity

A. K. Karakka Kal et al., *Advanced Extraction Techniques for Drug and Toxin Detection*, https://doi.org/10.1007/978-3-032-17335-5_4

4.1 Prohibited Substances and Methods

Curbing prohibited substances and methods from the backbone of international sports governance, serving as essential tools to safeguard fairness, protect the health of athletes, and maintain the integrity of competition [1, 2]. The issue of doping is not new; it traces back to the 1960s, when growing concerns over the misuse of stimulants first drew global attention. The IOC responded by introducing medical controls at the 1968 Winter Games, marking the beginning of structured anti-doping regulation [3, 4]. The establishment of WADA in 1999 was a milestone, followed by the adoption of the World Anti-Doping Code in 2004. Together, these initiatives harmonized fragmented rules into a unified framework and mandated the annual publication of the Prohibited List a cornerstone in the fight to ensure safe and fair competition [5].

The Prohibited List is updated every year to reflect new scientific findings and emerging threats [4, 6]. Substances and methods are included if they meet at least two of three criteria: They enhance performance, pose health risks, or undermine the spirit of sport [7, 8]. The list is divided into categories that apply universally (in- and out-of-competition), those restricted to in-competition use, and those prohibited only in certain sports.

4.1.1 Prohibited Substances

Prohibited substances are classified into groups (S0–S9), methods into categories (M1–M3), and additional sport-specific restrictions such as β-blockers fall under category P1. The WADA organizes prohibited substances into specific categories based on their physiological effects and potential risks [5, 9]. Table 4.1 outlines the major categories of prohibited substances as defined in the WADA Prohibited List.

Different classes of prohibited substances exert their effects in unique ways [5, 6]. Anabolic agents like testosterone and synthetic steroids stimulate muscle growth and recovery, while peptide hormones such as erythropoietin (EPO) increase oxygen transport through red blood cell production. Stimulants, including amphetamines and cocaine, boost alertness and aggression, whereas narcotics suppress pain, enabling longer exertion. Masking agents, such as diuretics and plasma expanders, are used to conceal doping practices, while advanced techniques like gene editing push the boundaries by manipulating genetic material to alter muscle performance or endurance.

Specified versus Non-specified Substances

- Specified substances are those more commonly used for reasons unrelated to performance enhancement, such as diuretics prescribed for medical conditions. If an athlete can demonstrate unintentional use, penalties may be reduced.

Table 4.1 Categories of prohibited substances as defined by WADA

Category	Prohibited substances/methods	Status (specified/ non-specified)	Prohibited times	Key examples
S0: Non-approved substances	Substances not approved for human therapeutic use (e.g., veterinary drugs, designer drugs)	Specified	All times	BPC-157, 2,4-Dinitrophenol (DNP), ryanodine receptor stabilizers (e.g., S-107)
S1: Anabolic agents	Synthetic anabolic–androgenic steroids (AAS) and endogenous (natural) anabolic hormones promoting muscle growth and recovery	Non-specified	All times	Testosterone, nandrolone, clenbuterol, SARMs (e.g., ostarine, ligandrol)
S2: Peptide HORMONES AND GROWTH FACTORS	Substances affecting erythropoiesis, growth, or hormone regulation	Non-specified	All times	Erythropoietin (EPO), human growth hormone (hGH), HIF activators (e.g., roxadustat)
S3: Beta-2 agonists	Inhalers for asthma; systemic use prohibited	Specified (inhaled)/ Non-specified (systemic)	All times (systemic)	Terbutaline (requires exemption)
S4: Hormone and metabolic modulators	Drugs altering hormone function or metabolism	Non-specified	All times	Tamoxifen, aromatase inhibitors (e.g., anastrozole), myostatin inhibitors
S5: Diuretics and masking agents	Substances altering urine composition to hide doping	Specified	All times	Furosemide, hydrochlorothiazide, desmopressin
S6: Stimulants	Substances enhancing alertness or reducing fatigue	Specified/ Non-specified	In-competition	Cocaine, amphetamines, ephedrine (threshold limits apply)
S7: Narcotics	Pain-relieving drugs with addiction potential	Specified	In-competition	Morphine, fentanyl, oxycodone
S8: Cannabinoids	Natural or synthetic THC products	Specified	In-competition	Cannabis, hashish, synthetic cannabinoids (e.g., JWH-018)

(continued)

Table 4.1 (continued)

Category	Prohibited substances/methods	Status (specified/ non-specified)	Prohibited times	Key examples
S9: Glucocorticoids	Anti-inflammatory drugs prohibited via certain routes	Specified	In-competition (oral/ injectable)	Prednisone (oral), dexamethasone (injectable)
P1 β-Blockers	Propranolol, Atenolol		Substances Prohibited in Certain Sports (Archery, Shooting, Darts)	

- Non-specified substances, including anabolic steroids and other high-risk agents, are considered serious violations. A first offense typically results in a four-year ban.

Substances of Abuse: Drugs such as cocaine, heroin, and THC fall into this category. Because these are often linked to recreational misuse rather than performance enhancement, sanctions may be reduced to one to three months if the athlete completes a rehabilitation program.

Evidence of Performance Enhancement: Current research shows that only a small portion of prohibited categories about five out of twenty-three, including anabolic agents and stimulants have strong evidence of improving sprinting power or strength. For endurance performance, no clear benefits have been consistently demonstrated.

Medical Exemptions (TUEs): Athletes who require certain prohibited substances for legitimate health conditions, such as inhalers for asthma, must obtain a therapeutic use exemption (TUE). This ensures medical treatment without compromising fairness.

Annual Updates and Monitoring: The Prohibited List is reviewed and updated every year, effective from January 1. WADA also actively monitors new and emerging substances, such as selective androgen receptor modulators (SARMs), to determine whether they should be added to the list.

4.1.2 Prohibited Methods

Prohibited methods extend beyond chemical use [5, 8]. As outlined in Table 4.2, WADA classifies prohibited methods into specific categories to safeguard fairness and athlete health. Blood doping, which involves transfusing stored or donor blood to elevate hemoglobin levels, is one such example. Physical manipulations, like urine substitution or catheterization, are also banned as they compromise testing integrity.

Table 4.2 Categories of prohibited methods in sport

Category	Prohibited substances/ methods	Status (specified/ non-specified)	Prohibited times	Key examples
M1: Blood manipulation	Methods altering blood components for oxygen enhancement	Non-specified	All times	Blood doping, red blood cell transfusions
M2: Chemical tampering	Altering samples or using IV infusions (>100 mL/12h)	Non-specified	All times	Urine substitution, intravenous hydration beyond limits
M3: Gene cell and doping	Gene editing, Stem-cell enhancements	Non-specified	All times	EPO gene transfer

Even non-invasive techniques, such as misusing hypoxic chambers to mimic high-altitude training, fall under regulation when they grant athletes undue physiological advantages.

Major Modifications in 2025

- *Blood donation* is now permitted provided it is carried out in accredited medical centers, such as through apheresis programs.
- *Beta-blockers* have been removed from the Prohibited List in specific winter disciplines, namely skiing and snowboarding.
- *Formoterol dosing* guidelines were updated to allow administration every 12 h, while the maximum daily limit of 54 μg remains unchanged.

4.1.3 Scientific and Ethical Grounds for Prohibition

The rationale behind these prohibitions is multifaceted. First and foremost, they protect athlete safety. Many banned substances have severe side effects, including cardiovascular disease, infertility, and psychological harm. Secondly, they uphold the credibility of sport. When doping is allowed to prevail, competitions become contests of chemistry rather than skill and dedication. To counter this, WADA has introduced rigorous monitoring systems, including biological passports and surprise testing, to track athletes' health markers over time and discourage violations [5, 9].

Yet, despite these measures, challenges persist. The development of designer drugs intended to evade detection, and the complex ethical issues surrounding TUEs, continue to test regulatory frameworks. High-profile doping scandals, such as the systematic program in Russian athletics or the misuse of β-blockers in precision sports, highlight the ongoing battle between scientific innovation and regulatory control.

Criteria for Inclusion: A substance or method is prohibited if it meets at least two of three conditions [5, 7].

- It has the potential to enhance performance.
- It poses actual or potential health risks.
- It contradicts the spirit of sport and principles of fairness.

Newly Added Substances (2025)

- *S0*: Ryanodine receptor stabilizers (e.g., S-107, ARM210).
- *S4*: Elacestrant (an anti-estrogen) and MOTS-c.
- *S5*: Xipamide, a diuretic.

Therapeutic Use Exemptions (TUEs): Athletes requiring prohibited substances for genuine medical conditions such as insulin for diabetes must apply for a TUE. Applications are reviewed under strict medical and ethical guidelines by independent panels.

Monitoring Program: WADA continues to track emerging substances that are not yet prohibited but could be misused [5, 10].

Ultimately, the prohibition of substances and methods embodies a collective commitment to celebrate authentic human achievement. Sport is meant to showcase discipline, resilience, and natural talent rather than artificial enhancement. As scientific and technological advances continue, the balance between innovation and fairness must be carefully maintained. This requires vigilance, transparency, and cooperation on a global scale. The updates in the 2025 Prohibited List reflect this evolving effort, providing refined guidelines that reinforce the principles of safety, fairness, and integrity in modern sport.

Annual Review Process: Updates to the Prohibited List are made through a structured process involving:

- Consultation with scientific experts, law enforcement, and pharmaceutical researchers.
- Circulation of draft versions for review by anti-doping organizations.
- Final approval by WADA's Executive Committee in September of each year.

4.2 Mechanisms and Performance Effects

Prohibited substances act through diverse biological mechanisms, each designed to enhance performance but often at the expense of athlete health and the principles of fair play [11]. Some substances work by promoting muscle anabolism or boosting oxygen delivery, while others stimulate the central nervous system or manipulate physiological pathways. Despite these performance-enhancing effects, their use frequently introduces serious medical risks and undermines the integrity of competition.

A key mechanism involves *neurotransmitter modulation*, where drugs influence brain chemistry by altering dopamine, serotonin, or GABA pathways. Such changes can affect mood, cognition, and physical output. Similarly, *receptor-specific actions* are seen in beta-blockers, which target β-adrenergic receptors to reduce anxiety and tremors, or opioids, which act on μ-opioid receptors to blunt pain [11]. While these mechanisms can provide short-term benefits, they also carry risks such as dangerously slow heart rates or respiratory depression.

Enzyme inhibition is another pathway, as with non-steroidal anti-inflammatory drugs (NSAIDs) or ACE inhibitors [11]. These agents block enzymes like COX or ACE to manage inflammation or blood pressure but may interfere with protective functions, such as gastric lining integrity or electrolyte regulation. The risk of dependence is particularly concerning for stimulants and benzodiazepines, which can rapidly lead to addiction due to neuroadaptive changes in the brain.

Furthermore, genetic variability plays a significant role in drug responses. Polymorphisms in enzymes such as CYP450 can alter how individuals metabolize substances, leading to unpredictable differences in efficacy and toxicity. This has direct clinical implications, particularly for personalized dosing, where genetic testing (e.g., CYP2D6 in codeine metabolism) can guide safer and more effective treatment strategies.

These complexities highlight the need for careful monitoring of drug interactions, especially in athletes exposed to multiple therapies. Emerging fields such as biologics and chronopharmacology which optimizes drug timing to body rhythms promise greater precision but also demand strict oversight. Table 4.3 provides a structured summary of major doping classes, their mechanisms of action, and documented effects on performance, integrating both pharmacological principles and clinical insights.

4.3 Antipsychotic Drugs

Antipsychotic medications, often referred to as neuroleptics, represent a fundamental class of drugs in the management of psychotic disorders as well as a variety of psychiatric and neurological conditions [12–14]. Since their introduction in the 1950s, these agents have undergone significant evolution. The earliest forms, known as first-generation or "typical" antipsychotics, primarily function by blocking dopamine D_2 receptors. While effective against positive symptoms such as hallucinations and delusions, they were associated with high rates of movement-related side effects. The development of second-generation or "atypical" antipsychotics marked a major advancement, as these agents act on not only dopamine receptors but also antagonize serotonin 5-HT_2A and other receptor systems, offering broader symptom control with improved tolerability.

Table 4.3 Summary of key doping classes, their primary mechanisms of action, and documented performance effects

Drug class	Mechanism of action	Performance effects	Key examples
Antipsychotics	Block dopamine D2 receptors and serotonin receptors (e.g., 5-HT2A). Atypical agents (e.g., clozapine) also modulate muscarinic M4 receptors	Reduces psychosis and agitation; may cause motor side effects (e.g., tardive dyskinesia) or metabolic issues (weight gain)	Clozapine, Risperidone, Haloperidol
Stimulants	Antagonize adenosine receptors (A1, A2A) and inhibit phosphodiesterase, increasing dopamine/norepinephrine release	Enhances alertness, focus, and endurance; chronic use risks tachycardia, anxiety, or dependence	Caffeine, Amphetamines, Modafinil
Opioids	Activate μ-opioid receptors, inhibiting pain signals and increasing dopamine in the nucleus accumbens	Provides potent analgesia and euphoria; suppresses respiratory drive and causes sedation	Morphine, Fentanyl, Oxycodone
Benzodiazepines	Enhance GABA-A receptor activity, increasing chloride influx and neuronal inhibition	Reduces anxiety, seizures, and muscle spasms; causes sedation, cognitive impairment, and dependence	Diazepam, Alprazolam, Lorazepam
Beta-Blockers	Block β-adrenergic receptors, reducing sympathetic nervous system activity	Lowers heart rate and blood pressure; mitigates anxiety but may impair exercise tolerance	Propranolol, Metoprolol, Atenolol
NSAIDs	Inhibit cyclooxygenase (COX) enzymes, reducing prostaglandin synthesis	Alleviates pain and inflammation; risks gastrointestinal ulcers and impaired blood clotting	Ibuprofen, Naproxen, Aspirin
Cephalosporins	Disrupt bacterial cell wall synthesis via β-lactam ring binding to penicillin-binding proteins	Treats infections (e.g., respiratory, urinary); overuse risks antibiotic resistance	Ceftriaxone, Cefazolin, Cephalexin
SSRIs/SNRIs	Inhibit serotonin/norepinephrine reuptake, increasing synaptic neurotransmitter levels	Improves mood and anxiety; initial use may cause agitation, insomnia, or sexual dysfunction	Fluoxetine, Sertraline, Venlafaxine

(continued)

Table 4.3 (continued)

Drug class	Mechanism of action	Performance effects	Key examples
Cannabinoids	Activate CB1/CB2 receptors, modulating dopamine release and glutamate/GABA signaling	Reduces pain and nausea; impairs motor coordination and memory	THC, CBD, Synthetic Cannabinoids
Antihypertensives (ACE Inhibitors)	Inhibit angiotensin-converting enzyme (ACE), reducing angiotensin II and vasoconstriction	Lowers blood pressure; may cause cough or hyperkalemia	Lisinopril, Enalapril, Ramipril

4.3.1 Classification and Examples

First-Generation (Typical) Antipsychotics: These drugs are potent D_2 receptor antagonists, effective in reducing positive symptoms of psychosis but with a considerable risk of extrapyramidal symptoms (EPS). The chemical structures of selected first-generation antipsychotics are presented in Fig. 4.1.

- Chlorpromazine, the earliest phenothiazine, noted for its sedative and anticholinergic effects.
- Haloperidol, a butyrophenone with high dopamine receptor affinity, available in oral and long-acting injectable forms.
- Other notable agents include fluphenazine, perphenazine, thiothixene, and pimozide, the latter also used in Tourette's disorder.

Second-Generation (Atypical) Antipsychotics: These agents combine dopamine and serotonin receptor antagonism, offering benefits for both positive and negative symptoms with fewer motor side effects. However, they present higher risks of metabolic complications. The chemical structures of selected second-generation antipsychotics are presented in Fig. 4.2.

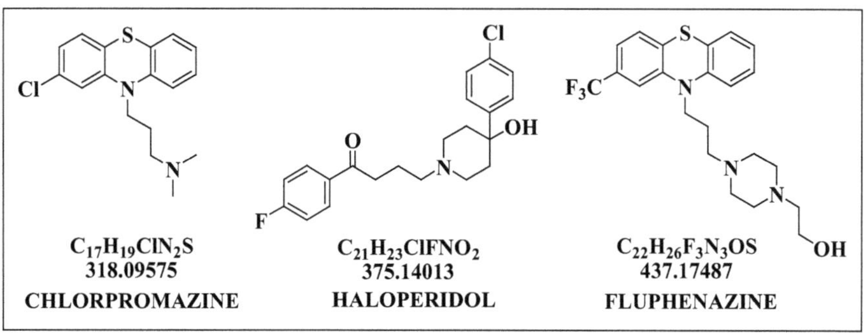

Fig. 4.1 Chemical structure of selected first-generation antipsychotics

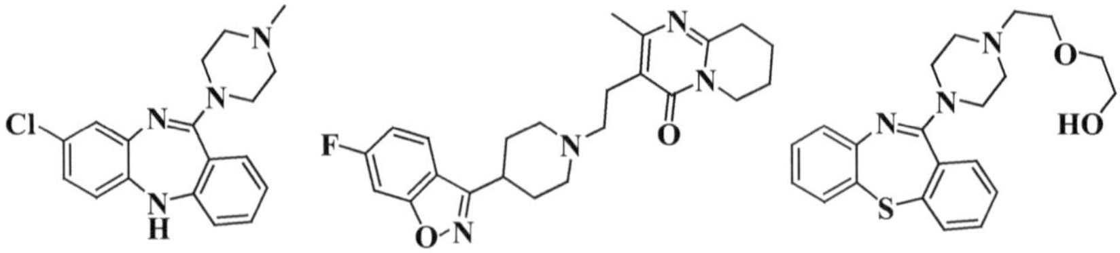

$C_{18}H_{19}ClN_4$	$C_{23}H_{27}FN_4O_2$	$C_{21}H_{25}N_3O_2S$
326.12982	410.2118	383.16675
CLOZAPINE	**RISPERIDONE**	**QUETIAPINE**

Fig. 4.2 Chemical structure of selected second-generation antipsychotics

Key examples include the following:

- Clozapine, highly effective in treatment-resistant schizophrenia but requiring strict blood monitoring due to the risk of agranulocytosis.
- Widely used drugs such as risperidone, olanzapine, quetiapine, aripiprazole, ziprasidone, and paliperidone.
- Newer additions include lurasidone, asenapine, brexpiprazole, and cariprazine, designed to improve efficacy and tolerability.

4.3.2 Mechanisms of Action

- **D_2 Receptor Antagonism**: Reduces overactivity in mesolimbic dopamine pathways, alleviating hallucinations and delusions, but can cause motor side effects via nigrostriatal blockade.
- **5-HT_2A Receptor Antagonism**: Typical of atypical antipsychotics, this mechanism helps mitigate negative and cognitive symptoms while reducing extrapyramidal symptoms (EPS).
- **Additional Targets**: Many agents act on histamine, muscarinic, and adrenergic receptors, leading to sedation, anticholinergic effects, or orthostatic hypotension.

4.3.3 Clinical Applications

Antipsychotics are central in the treatment of schizophrenia and schizoaffective disorder, serving as first-line agents in both acute and maintenance phases. While both generations control positive symptoms effectively, atypicals are preferred for addressing negative and cognitive features. In bipolar disorder, particularly during manic or mixed episodes, drugs such as quetiapine and olanzapine are FDA-approved and often used in combination with mood stabilizers. Some atypicals, such as quetiapine and aripiprazole, are also prescribed as augmentation strategies in treatment-resistant depression. In addition, low-dose antipsychotics may be used for behavioral disturbances in dementia or autism, though caution is required due to safety concerns and regulatory warnings.

4.3.4 *Adverse Effects and Monitoring*

Movement Disorders: First-generation agents carry a high risk of acute dystonia, parkinsonism, and akathisia, with tardive dyskinesia developing in long-term use. Although less common with atypicals, these effects may still occur at high doses.

Metabolic and Cardiovascular Risks: Weight gain, diabetes, and lipid abnormalities are most pronounced with clozapine and olanzapine. Cardiac monitoring is essential for agents like ziprasidone or iloperidone due to QT prolongation.

Hematologic and other Concerns: Clozapine use requires routine complete blood counts due to the risk of agranulocytosis. Additional side effects include anticholinergic symptoms such as dry mouth and constipation, as well as prolactin elevation with risperidone and paliperidone.

4.3.5 *Emerging Directions*

Recent research emphasizes pharmacogenetics, where genetic profiling may help optimize dosing and drug selection, improving safety and efficacy. Novel drug classes, including glutamatergic modulators, trace-amine receptor agonists, and biased dopamine ligands, are under investigation to provide effective symptom relief with fewer side effects. Long-acting injectables (LAIs) represent another important development, offering improved adherence and reducing relapse rates in chronic conditions like schizophrenia and bipolar disorder. Table 4.4 provides a comprehensive summary of commonly encountered antipsychotic drugs, outlining their pharmacological classification, mechanism of action, and significance in forensic and anti-doping investigations.

Table 4.4 Summary of common antipsychotic drugs: classification and forensic relevance

Generic name	Chemical class	Notable characteristics
First-Generation Antipsychotics		
Chlorpromazine	Phenothiazine	First antipsychotic developed; known for sedative effects; higher risk of extrapyramidal symptoms (EPS)
Haloperidol	Butyrophenone	High-potency; effective for acute psychosis; available in long-acting injectable form
Fluphenazine	Phenothiazine	High-potency; long-acting injectable available; used for chronic schizophrenia
Perphenazine	Phenothiazine	Medium-potency; balances efficacy and side effect profile
Thioridazine	Phenothiazine	Low-potency; higher anticholinergic side effects; risk of cardiac arrhythmias
Trifluoperazine	Phenothiazine	High-potency; used for schizophrenia and anxiety disorders
Loxapine	Dibenzoxazepine	Medium-potency; has properties similar to atypical antipsychotics; available in inhalation form for acute agitation
Pimozide	Diphenylbutylpiperidine	Used primarily for Tourette's disorder; requires cardiac monitoring due to QT prolongation risk
Thiothixene	Thioxanthene	High-potency; used for schizophrenia; risk of EPS
Zuclopenthixol	Thioxanthene	Available in oral and long-acting injectable forms; used for acute and chronic psychoses
Second-Generation Antipsychotics		
Clozapine	Dibenzodiazepine	Effective for treatment-resistant schizophrenia; risk of agranulocytosis; requires regular blood monitoring
Risperidone	Benzisoxazole	Broad efficacy; available in long-acting injectable form; may cause prolactin elevation
Olanzapine	Thienobenzodiazepine	Effective for schizophrenia and bipolar disorder; associated with significant weight gain and metabolic effects
Quetiapine	Dibenzothiazepine	Used for schizophrenia, bipolar disorder, and as adjunct in depression; sedating properties
Aripiprazole	Quinolinone	Partial dopamine agonist; lower risk of weight gain; available in long-acting injectable form
Ziprasidone	Benzisothiazole	Lower risk of weight gain; requires food intake for proper absorption; risk of QT prolongation
Paliperidone	Benzisoxazole	Active metabolite of risperidone; available in long-acting injectable forms; used for schizophrenia and schizoaffective disorder

(continued)

Table 4.4 (continued)

Generic name	Chemical class	Notable characteristics
Lurasidone	Benzisothiazole	Used for schizophrenia and bipolar depression; favorable metabolic profile; must be taken with food
Asenapine	Dibenzoxepinopyrrole	Sublingual administration; used for schizophrenia and bipolar disorder; minimal weight gain
Iloperidone	Piperidinyl-benzisoxazole	Used for schizophrenia; requires gradual dose titration due to orthostatic hypotension risk
Brexpiprazole	Quinolinone	Similar to aripiprazole; used for schizophrenia and as adjunct in major depressive disorder; lower risk of akathisia
Cariprazine	Piperazine	Partial agonist at dopamine receptors; used for schizophrenia and bipolar disorder; long half-life
Amisulpride	Benzamide	Selective dopamine antagonist; used for schizophrenia; not approved in all countries
Lumateperone	Butyrophenone derivative	Newer agent; favorable side effect profile; used for schizophrenia and bipolar depression
Sertindole	Phenylindole derivative	Withdrawn in some countries due to cardiac concerns; used for schizophrenia
Zotepine	Dibenzothiazepine	Used for schizophrenia; not widely available globally

4.4 Stimulant Drugs

Stimulant drugs represent a broad and diverse class of compounds known for their ability to enhance central nervous system (CNS) activity [15–17]. These agents are widely utilized in clinical practice for conditions such as attention-deficit/hyperactivity disorder (ADHD) and narcolepsy, while also carrying substantial potential for illicit use, exemplified by substances like cocaine and methamphetamine. Stimulants primarily act by modulating monoaminergic neurotransmission, affecting dopamine, norepinephrine, and, to varying degrees, serotonin. This pharmacological action results in heightened alertness, energy, focus, and often a sense of euphoria. Clinically, stimulants provide meaningful benefits by improving attention, reducing hyperactivity, and promoting wakefulness; however, their misuse can lead to significant health consequences, including cardiovascular complications, addiction, neuropsychiatric disturbances, and neurotoxicity. Therefore, responsible prescribing, appropriate dosing, and careful monitoring are essential to balance therapeutic benefits against potential risks. Figure 4.3 illustrates the chemical structures of key stimulant drugs frequently encountered in forensic toxicology and doping control.

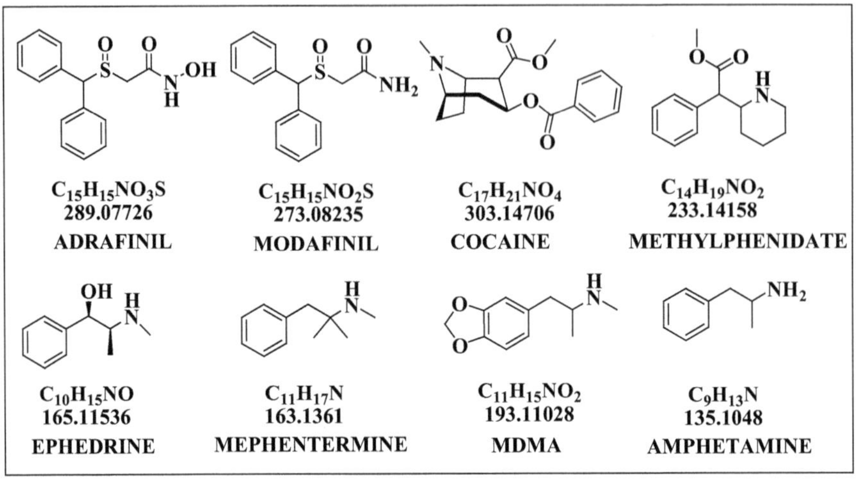

Fig. 4.3 Representative chemical structures of major stimulant drugs commonly detected in forensic and anti-doping analyses

4.4.1 Classification

Therapeutic Stimulants

- Amphetamines (e.g., Adderall, mixed amphetamine salts) increase the release of dopamine and norepinephrine, serving as a first-line therapy for ADHD and narcolepsy.
- Methylphenidate (Ritalin, Concerta) inhibits the reuptake of dopamine and norepinephrine, enhancing synaptic catecholamine levels; it is also widely used in ADHD and narcolepsy.
- Modafinil/Armodafinil (Provigil, Nuvigil) promote wakefulness by modulating hypothalamic orexin/hypocretin pathways and are associated with relatively low abuse potential.

Illicit and Semi-Licit Stimulants

- Cocaine acts by blocking monoamine reuptake and sodium channels, producing intense euphoria and high addictive potential.

Methamphetamine triggers massive monoamine release, posing significant risks of abuse and long-term neurotoxicity.

Ephedrine and Pseudoephedrine (Bronkaid, Sudafed) function as α/β-adrenergic agonists used in decongestants and also serve as precursors for methamphetamine synthesis.

4.4.2 Mechanisms of Action

Stimulants operate through several distinct mechanisms:

Monoamine Release: Amphetamines and methamphetamine reverse the function of dopamine and norepinephrine transporters (DAT/NET), flooding synapses with neurotransmitters.

Reuptake Inhibition: Methylphenidate and cocaine prolong neurotransmitter action by blocking DAT and NET (with cocaine also affecting serotonin transporters to a lesser extent).

Wakefulness Promotion: Modafinil exerts its effects through orexinergic, histaminergic, and glutamatergic pathways, rather than direct increases in monoamines.

Table 4.5 provides an overview of key stimulant drugs, highlighting their mechanisms of action, pharmacological effects, and therapeutic or illicit applications.

4.4.3 Clinical Uses

Stimulants are highly effective in several clinical contexts:

ADHD: They enhance prefrontal cortex catecholamine signaling, improving attention, reducing impulsivity, and controlling hyperactivity.

Narcolepsy and Shift-Work Sleep Disorder: Wake-promoting stimulants maintain daytime alertness and reduce excessive sleepiness.

Off-Label and Cognitive Enhancement: While some individuals misuse stimulants for study or cognitive enhancement, the benefits are modest and accompanied by heightened risks of cardiovascular events and dependence.

4.4.4 Adverse Effects and Risks

Potential adverse outcomes include the following:

Cardiovascular: Hypertension, tachycardia, arrhythmias, and stroke, especially at high doses or with illicit use.

Neuropsychiatric: Anxiety, insomnia, agitation, and psychosis, particularly with chronic or high-dose exposure.

Dependence and Withdrawal: Physical and psychological dependence can occur, with withdrawal symptoms including fatigue, depression, and dysphoria.

Table 4.5 Summary of major stimulant drugs, their pharmacological mechanisms, and clinical/forensic relevance

Generic name	Chemical class	Notable remarks
Adrafinil	Eugeroic	Prodrug of modafinil; stimulates wakefulness)
Adrenaline (Epinephrine)	Catecholamine	Local use permitted; systemic administration banned in-competition
Amfepramone (Diethylpropion)	Phenethylamine	Anorectic agent with stimulant effects; often abused for weight loss
Amiphenazole	Respiratory stimulant	Historically used to reverse barbiturate overdose
Amphetamine	Phenethylamine	Prototype CNS stimulant; includes d-amphetamine, l-amphetamine, mixed salts
Armodafinil	Eugeroic	R-enantiomer of modafinil; wakefulness promoter
Benfluorex	Aminorex derivative	Withdrawn due to cardiovascular risk
Benzphetamine	Phenethylamine	Anorectic stimulant; metabolized to amphetamine
Benzylpiperazine (BZP)	Piperazine	Recreational stimulant; chemically unrelated to amphetamines
Bromantane	Adamantane derivative	Adaptogen with stimulant properties; used in Russia
Cathine (Norpseudoephedrine)	Phenethylamine	Naturally occurring in khat; threshold 5 μg/mL in urine
Cathinone	β-Keto-phenethylamine	Active compound in khat; parent to many NPS
Clobenzorex	Amphetamine derivative	Anorectic; metabolized to chlorphentermine
Cocaine	Tropane alkaloid	Potent CNS stimulant; local anesthetic
Cropropamide	Respiratory stimulant	Formerly used for barbiturate reversal
Cyclazodone	Amphetamine analogue	Designer stimulant
Dimethylamphetamine	Phenethylamine	Structural analogue of methamphetamine
Ephedrine	Phenethylamine	Bronchodilator; nasal decongestant

(continued)

Table 4.5 (continued)

Generic name	Chemical class	Notable remarks
Etamivan	Respiratory stimulant	Formerly used to manage barbiturate overdose
Etilamphetamine	Phenethylamine	Prodrug of amphetamine
Famprofazone	Pyrazolone derivative	Analgesic with stimulant side effects
Fenbutrazate	Phenethylamine	Anorectic stimulant
Fencamfamine	Phenethylamine	Central stimulant
Fenfluramine	Phenethylamine derivative	Appetite suppressant; withdrawn due to cardiac risks
Fenproporex	Amphetamine derivative	Anorectic; precursor to amphetamine
Furfenorex	Amphetamine derivative	Anorectic stimulant
Heptaminol	Amino alcohol	Cardiac stimulant
Isometheptene	Aminoalkane	Vasoconstrictor; migraine treatment
Levo-methamphetamine	Phenethylamine	l-Isomer of methamphetamine
Meclofenoxate	Choline ester	Nootropic with stimulant effects
Mefenorex	Amphetamine derivative	Anorectic; precursor to amphetamine
Mephentermine	Phenethylamine	Cardiac stimulant
Mesocarb	Amphetamine analogue	Investigational stimulant
Methamphetamine	Phenethylamine	Highly potent; includes d- and l-isomers
4-Methylamphetamine	Phenethylamine	Structural analogue of amphetamine
Methylenedioxyamphetamine (MDA)	Phenethylamine	Amphetamine analogue; psychoactive
Methylenedioxymethamphetamine (MDMA)	Phenethylamine	"Ecstasy"; empathogen, stimulant
Methylephedrine	Phenethylamine	Stimulant metabolite of ephedrine
Methylphenidate	Piperidine derivative	ADHD medication; potent CNS stimulant
Modafinil	Eugeroic	Wakefulness promoter; low abuse potential
Nikethamide	Respiratory stimulant	Formerly used to treat barbiturate overdose

(continued)

Table 4.5 (continued)

Generic name	Chemical class	Notable remarks
Norfenefrine	Phenethylamine	Similar to phenylephrine; vasoconstrictor
Norfenfluramine	Phenethylamine	Metabolite of fenfluramine
Octopamine	Trace amine	Endogenous in some species; stimulant properties
Ortetamine	Phenethylamine	Stimulant; structural analogue of amphetamine
Oxilofrine	Phenethylamine	Stimulant used as a nasal decongestant
Parahydroxyamphetamine	Phenethylamine	Metabolite of amphetamine
Pemoline	Oxazolidinone	ADHD medication; liver toxicity risk
Pentetrazol (Metrazol)	Piperazine derivative	Convulsant/stimulant; historically used in shock therapy
Phendimetrazine	Phenethylamine	Anorectic stimulant
Phenmetrazine	Phenethylamine	Anorectic; popular in mid-twentieth century
Phentermine	Phenethylamine	Common weight-loss drug; structurally related to amphetamine
4-Phenylpiracetam (Carphedon)	Racetam	Cognitive enhancer with stimulant effects
Prenylamine	Calcium channel blocker	Stimulant side effects
Prolintane	Piperidine derivative	CNS stimulant
Propylhexedrine	Cycloalkylamine	Over-the-counter nasal decongestant
Selegiline (Deprenyl)	MAO-B inhibitor	Parkinson's medication; stimulant at higher doses
Sibutramine	ATP-citrate lyase inhibitor	Appetite suppressant; withdrawn for CV risk
Strychnine	Alkaloid	Poisonous stimulant at low doses
Tuaminoheptane	Amino alcohol	Nasal decongestant; stimulant properties
Midodrine	Imidazoline derivative	Added to S6 in 2025; used for orthostatic hypotension
Tesofensine	Triple monoamine reuptake inhibitor	Added to S6 in 2025; investigated for obesity

Neurotoxicity: Methamphetamine use can damage dopaminergic neurons, potentially leading to long-term cognitive deficits.

Key Notes

Scope of Prohibition: All optical isomers and salts of the stimulants listed are prohibited in-competition under Section S6 of the WADA Prohibited List.

2025 Updates: Midodrine and Tesofensine have been added to the S6 category in WADA's 2025 update.

4.5 Opioid Drugs

Opioid drugs constitute a broad and diverse class of compounds including natural, semi-synthetic, and fully synthetic agents renowned for their potent analgesic effects [18, 19]. These medications relieve pain primarily by activating opioid receptors located throughout the central and peripheral nervous systems. Clinically, opioids are indispensable for managing moderate-to-severe pain, particularly in postoperative, cancer-related, or palliative settings. However, their use carries substantial risks, including tolerance, physical dependence, overdose, and a range of systemic adverse effects. Over the past two decades, a combination of increased prescribing and the widespread availability of potent illicit synthetics has precipitated a public health crisis, with escalating rates of overdose-related fatalities. Modern medical practice emphasizes balanced pain management strategies that prioritize careful patient selection, dose minimization, and comprehensive approaches to reduce the potential for misuse and dependence. The chemical structures of representative opioid compounds are illustrated in Fig. 4.4.

4.5.1 Classification and Examples

Opioids can be classified based on their origin and chemical structure:

Natural Opioids (Opiates): Morphine and codeine are examples of alkaloids that extracted directly from the opium poppy plant. They undergo first-pass metabolism, producing active or sometimes toxic metabolites.

Semi-synthetic Opioids: These are chemically modified derivatives of natural opiates designed to improve potency, bioavailability, or pharmacokinetics. Examples include hydrocodone, oxycodone, hydromorphone, and oxymorphone.

Synthetic Opioids: Fully synthetic compounds, such as fentanyl, methadone, meperidine, buprenorphine, tramadol, and tapentadol, display varying receptor affinities and half-lives, providing a range of therapeutic options.

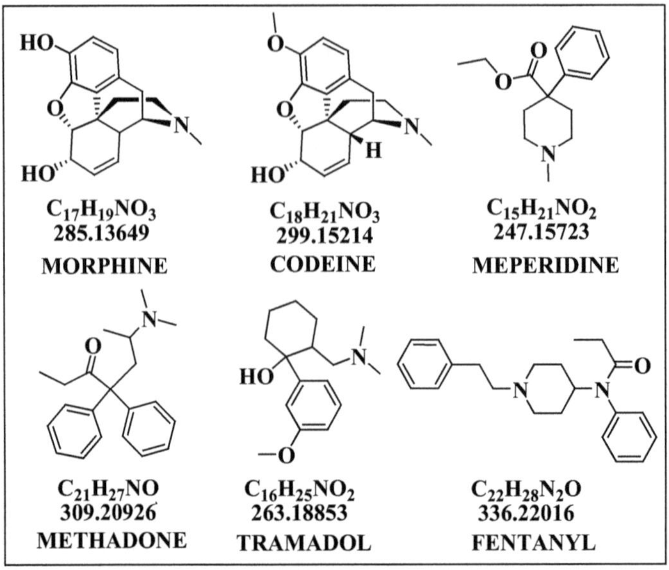

Fig. 4.4 Chemical structures of major opioid compounds

4.5.2 *Mechanisms of Action*

Opioids exert their analgesic effects primarily through μ-opioid receptor agonism. At the spinal level, they inhibit nociceptive transmission by reducing presynaptic neurotransmitter release and hyperpolarizing postsynaptic neurons. In addition, opioids activate descending inhibitory pathways from the brainstem, further dampening pain signals. Their action on limbic system structures contributes to euphoria, which, while beneficial for perceived pain relief, also underlies the high potential for misuse and addiction.

4.5.3 *Clinical Uses*

Acute Pain: Opioids are used short-term for postoperative or injury-related pain, often employing immediate-release formulations.

Chronic Non-cancer Pain: Long-term use is reserved for refractory cases and guided by strict protocols. Current guidelines, such as those from the CDC (2022), recommend the lowest effective dose for the shortest duration feasible.

Cancer Pain and Palliative Care: Opioids remain central in managing severe pain and dyspnea, with fewer restrictions on dose titration.

Opioid-Use Disorder (OUD) Treatment: Methadone and buprenorphine maintenance therapies reduce illicit opioid use and associated mortality, forming a cornerstone of evidence-based addiction management.

4.5.4 Adverse Effects and Risks

Opioids carry a broad spectrum of potential adverse effects:

Respiratory Depression: Dose-dependent suppression of brainstem respiratory centers is the leading cause of fatal overdose.

Tolerance and Physical Dependence: Repeated exposure diminishes analgesic effects, and abrupt cessation triggers withdrawal symptoms.

Gastrointestinal Effects: Constipation is nearly universal and often requires prophylactic management.

Neuroendocrine Effects: Chronic use can lead to hypogonadism and osteoporosis.

Cognitive and Psychomotor Impairment: These effects increase the risk of falls, accidents, and impaired driving.

4.5.5 Public Health Impact and Harm-Reduction Strategies

Between 1999 and 2012, prescription opioid sales in the U.S. quadrupled, accompanied by a more-than-threefold increase in opioid-related fatal overdoses. In response, multiple strategies have been implemented to mitigate harm:

Prescription Monitoring Programs (PMPs): Track opioid dispensing to prevent "doctor shopping."

Risk Assessment Tools: Evaluate patient history, comorbidities, and other factors before prescribing opioids.

Opioid Stewardship and Clinical Guidelines: Adopt CDC and professional society recommendations for initiation, titration, and tapering.

Access to Naloxone: Distribution of Naloxone to communities allows reversal of accidental overdoses.

Multimodal Pain Management: Emphasis on non-opioid analgesics, physical therapy, and behavioral interventions to reduce opioid reliance.

Table 4.6 summarizes the key opioid drugs and their important characteristics. In conclusion, opioids remain essential for the relief of severe pain, but their use demands careful adherence to evidence-based guidelines. Judicious prescribing, patient education, and integrated public health measures are crucial to maximize therapeutic benefit while minimizing the risks of dependence, overdose, and broader societal harm.

Table 4.6 Key opioid drugs and their characteristics

Generic name	Chemical class	Notable characteristics
Morphine	Phenanthrene alkaloid (natural)	Gold-standard opioid analgesic; significant first-pass metabolism to active metabolites; high risk of constipation and tolerance
Codeine	Phenanthrene alkaloid (natural)	Prodrug converted to morphine via CYP2D6; mild analgesic and antitussive; efficacy varies with genetic polymorphisms
Hydrocodone	Phenanthrene derivative	Semi-synthetic; often combined with acetaminophen; intermediate potency; metabolized to hydromorphone
Oxycodone	Phenanthrene derivative	Semi-synthetic; high oral bioavailability; extended-release formulations available; significant abuse potential
Hydromorphone	Morphinan derivative	Semi-synthetic; ~5–7 × potency of morphine; available in immediate and ER forms; less histamine release
Oxymorphone	Morphinan derivative	Semi-synthetic; ~10 × potency of morphine; high risk of respiratory depression; ER formulations withdrawn in some regions
Fentanyl	Phenylpiperidine (synthetic)	Ultra-potent (~100 × morphine); transdermal patches and lozenges; rapid onset; high risk of overdose
Methadone	Diphenylheptane (synthetic)	Long-acting; NMDA antagonist properties; used in opioid-use disorder; risk of QT prolongation
Meperidine	Phenylpiperidine (synthetic)	Short-acting; metabolite normeperidine causes neurotoxicity (seizures); limited clinical use
Buprenorphine	Thebaine derivative (partial μ-agonist)	Ceiling effect on respiratory depression; used for opioid-use disorder; high affinity for μ-receptors
Tramadol	Aminocyclohexanol (atypical)	Weak μ-agonist plus SNRI; risk of serotonin syndrome; lower potency; dual mechanism
Tapentadol	Phenylropanol derivative	μ-agonist plus NRI; similar efficacy to oxycodone with fewer GI side effects; lower CYP interactions
Sufentanil	Phenylpiperidine (synthetic)	~5–10 × potency of fentanyl; used in anesthesia; very rapid onset and short duration
Alfentanil	Phenylpiperidine (synthetic)	Rapid onset, short duration; used for brief procedures; less potent than fentanyl

4.6 Benzodiazepines Drugs

Benzodiazepines, commonly referred to as “benzos”” are a class of psychoactive medications that enhance inhibitory neurotransmission mediated by γ-aminobutyric acid (GABA) within the central nervous system [20, 21]. By potentiating GABA activity, these drugs produce sedative, anxiolytic, anticonvulsant, and muscle-relaxant effects. First introduced in the 1960s as safer alternatives to barbiturates, benzodiazepines rapidly became foundational treatments for anxiety disorders, insomnia, seizure disorders, and alcohol-withdrawal symptoms. While their short-term efficacy is well established, prolonged or excessive use carries significant risks, including tolerance, physical dependence, cognitive decline, and in severe cases, life-threatening respiratory depression. These potential hazards necessitate careful prescribing practices, ongoing patient monitoring, and consideration of non-benzodiazepine alternatives where appropriate. The chemical structures of important benzodiazepines are presented in Fig. 4.5.

4.6.1 Mechanism of Action

Benzodiazepines act through allosteric modulation of the $GABA_A$ receptor, binding at the interface between the α and γ subunits. This binding increases the frequency

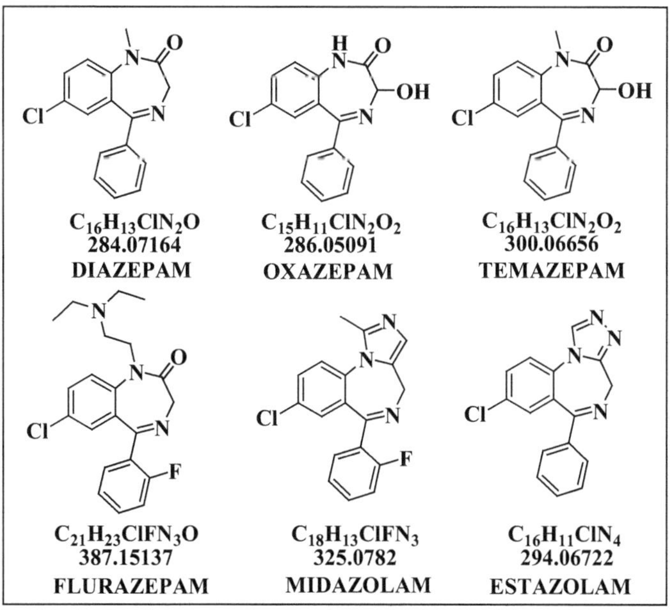

Fig. 4.5 Chemical structure of important benzodiazepines

of chloride-channel opening in the presence of GABA, hyperpolarizing neuronal membranes and reducing neuronal excitability. The enhancement of endogenous GABA activity underlies their sedative-hypnotic, anxiolytic, anticonvulsant, and muscle-relaxant effects. Variations in receptor-subtype selectivity, such as preference for α_1 versus $\alpha_2/_3$ subunits, contribute to differences in hypnotic versus anxiolytic potency among individual benzodiazepines. The key benzodiazepine agents and their associated clinical features are presented in Table 4.7 for reference.

4.6.2 Clinical Indications

Benzodiazepines are employed across several therapeutic contexts:

Anxiety Disorders: Used as first-line or adjunctive therapy for generalized anxiety disorder, panic disorder, and social anxiety, typically for short-term periods to reduce the risk of dependence.

Insomnia: Rapid-onset agents, such as temazepam and triazolam, facilitate sleep initiation; their use is limited to brief courses to prevent rebound insomnia.

Seizure Control: Diazepam and lorazepam are critical in emergency management of status epilepticus, while certain cases of refractory epilepsy may require chronic administration.

Alcohol Withdrawal: Agents like chlordiazepoxide mitigate withdrawal severity and reduce complications, including delirium tremens.

4.6.3 Adverse Effects and Safety Concerns

Common adverse effects include central nervous system depression, leading to drowsiness, dizziness, and impaired coordination, particularly in older adults. Chronic use may result in cognitive deficits, including memory impairments, attention difficulties, and potential long-term risks for dementia. Respiratory depression is rare when benzodiazepines are used alone but can be fatal when combined with opioids or alcohol. Paradoxical reactions, such as agitation, aggression, and disinhibition, may occur in a subset of patients.

4.6.4 Dependence, Withdrawal, and Misuse

Tolerance and physical dependence can develop within weeks of regular use. Abrupt discontinuation may trigger rebound anxiety, insomnia, or seizures. Withdrawal syndromes can include tremor, tachycardia, anxiety, and in severe cases, delirium

Table 4.7 Pharmacological profiles of benzodiazepine agents

Generic name	Chemical subclass	Notable characteristics
Chlordiazepoxide	1,4-Benzodiazepine	First benzodiazepine marketed (1960); long half-life (5–30 h); used for anxiety and alcohol-withdrawal syndromes; active metabolites prolong effect
Diazepam	1,4-Benzodiazepine	High bioavailability; long half-life (20–100 h) with active desmethyldiazepam; versatile (anxiety, seizures, muscle spasm, procedural sedation); various formulations (oral, IV, rectal)
Oxazepam	3-Hydroxybenzodiazepine	Metabolite of diazepam; intermediate-half-life (6–24 h); minimal drug–drug interactions (no phase I metabolism); preferred in elderly and liver impairment
Lorazepam	3-Hydroxybenzodiazepine	Intermediate-acting (10–20 h); strong anxiolytic and anticonvulsant; no active metabolites; widely used for status epilepticus and preoperative sedation
Clonazepam	1,4-Benzodiazepine	Long-acting (30–40 h); high anticonvulsant potency; indicated for absence, myoclonic seizures, and panic disorders; risk of tolerance and dependence
Temazepam	1,4-Benzodiazepine	Intermediate onset and half-life (8–20 h); hypnotic for insomnia; lower next-day residual effects compared to long-acting agents; metabolite-free
Triazolam	Triazolobenzodiazepine	Ultra-short-acting (½–2 h); rapid sleep induction; high risk of rebound insomnia and amnesia; limited to short-term use (< 7 days)
Flurazepam	1,4-Benzodiazepine	Long-acting (40–250 h); effective for sleep maintenance; significant next-day sedation and accumulation; largely superseded by shorter-acting hypnotics
Midazolam	Imidazobenzodiazepine	Very short-acting (1–4 h); potent amnestic and sedative; IV or intranasal use for procedural sedation; high clearance; water-soluble prodrug form
Nitrazepam	1,4-Benzodiazepine	Long-acting (16–48 h); hypnotic and anticonvulsant; risk of residual sedation; used for severe insomnia and nocturnal myoclonus
Clobazam	1,5-Benzodiazepine	Lower sedative effects; long-acting (36–42 h); adjunctive therapy for Lennox-Gastaut syndrome; less ataxia compared to classical benzodiazepines
Bromazepam	1,4-Benzodiazepine	Intermediate-acting (20–30 h); anxiolytic with moderate sedation; slow onset; used for anxiety and panic disorders; lower abuse potential than alprazolam

(continued)

Table 4.7 (continued)

Generic name	Chemical subclass	Notable characteristics
Clorazepate	1,4-Benzodiazepine	Prodrug of desmethyldiazepam; very long-acting (> 60 h); used for anxiety, seizures, and alcohol withdrawal; gradual onset limits misuse but prolongs sedation
Estazolam	1,4-Benzodiazepine	Intermediate-acting (10–24 h); used for insomnia; less next-day sedation than flurazepam; moderate risk of tolerance with prolonged use
Quazepam	1,4-Benzodiazepine	Long-acting (39–100 h); selective for sleep maintenance; unique active metabolite profile; lower rebound insomnia

or convulsions, necessitating gradual tapering under medical supervision. Benzodiazepines with rapid onset, such as alprazolam or diazepam, carry higher misuse potential, which is further exacerbated when co-administered with opioids.

4.6.5 *Risk Mitigation and Alternatives*

To reduce risks, prescribing guidelines recommend limiting duration (typically 2–4 weeks), using the lowest effective dose, and conducting periodic reassessment. Non-benzodiazepine alternatives, including SSRIs, SNRIs, buspirone, and select anticonvulsants, are preferred for long-term anxiety management. For insomnia, "z-drugs" such as zolpidem and zaleplon offer hypnotic benefits with lower dependence potential. Psychotherapeutic interventions, including cognitive-behavioral therapy and mindfulness-based approaches, can further reduce reliance on pharmacological treatment.

In summary, benzodiazepines remain highly valuable for the acute management of anxiety, insomnia, seizures, and alcohol withdrawal. Their safe and effective use requires careful patient selection, clearly defined indications, judicious dosing, and strict time limits to optimize therapeutic benefit while minimizing the risks of tolerance, dependence, and cognitive harm.

4.7 Beta-Blockers

Beta-blockers, or β-adrenoceptor antagonists, are a diverse class of medications that mitigate sympathetic stimulation of the heart and blood vessels by competitively blocking β-adrenergic receptors [22]. By reducing heart rate, myocardial contractility, and renin secretion, these agents form a cornerstone in the management of

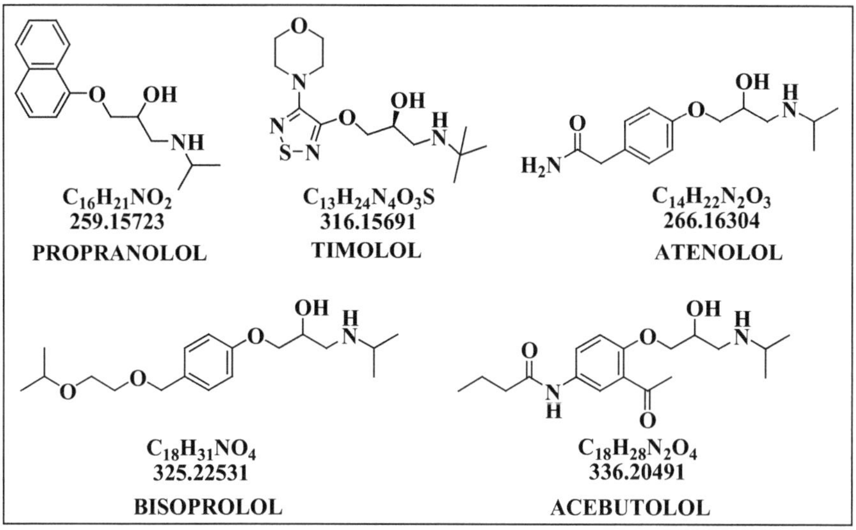

Fig. 4.6 Chemical structure of important beta-blockers

cardiovascular conditions such as hypertension, ischemic heart disease, heart failure, and various arrhythmias. Beyond their cardiovascular benefits, certain beta-blockers are also employed for non-cardiac indications, including migraine prevention, essential tremor, hyperthyroidism symptom control, and anxiety management [23, 24]. While highly effective, their use requires careful consideration, as nonselective agents can precipitate bronchospasm, metabolic disturbances, and fatigue. Selecting the appropriate agent and dose tailored to patient comorbidities is essential for maximizing benefits while minimizing risks. Figure 4.6 illustrates the chemical structures of commonly used beta-blockers.

4.7.1 Mechanism of Action

Beta-blockers function by binding competitively to β-adrenergic receptors, thereby preventing endogenous catecholamines epinephrine and norepinephrine from exerting their effects.

Cardioselective (β_1-selective) agents such as metoprolol and atenolol preferentially inhibit β_1 receptors in the heart and kidneys, reducing heart rate and renin secretion with minimal bronchial effects.

Nonselective agents, including propranolol and nadolol, block both β_1 and β_2 receptors, providing broader sympathetic inhibition but increasing the risk of bronchospasm and peripheral vasoconstriction.

Combined α/β-blockers such as labetalol and carvedilol additionally antagonize α_1 receptors, producing vasodilation that complements their β-blocking effects.

Table 4.8 Pharmacologic subclasses of β-blockers

Subclass	Examples	Features
Cardioselective β_1	Metoprolol, Atenolol, Bisoprolol	Reduced bronchial β_2 blockade; preferred in COPD/asthma; careful titration at higher doses as selectivity diminishes
Nonselective β_1/β_2	Propranolol, Nadolol, Timolol	Effective in portal hypertension and migraine; risk of bronchospasm and peripheral vasoconstriction; high CNS penetration (propranolol)
Partial agonists (ISA)	Acebutolol, Pindolol	Intrinsic sympathomimetic activity leads to less resting bradycardia; useful in patients prone to fatigue
Combined α/β-blockers	Labetalol, Carvedilol	Additional vasodilation via α_1 antagonism; beneficial in heart failure and hypertensive emergencies but may cause orthostatic hypotension
Ultra-short-acting β_1	Esmolol	IV use for rapid rate control; half-life ~9 min allows tight titration in acute care settings

Table 4.8 provides a summary of the pharmacologic subclasses of beta-blockers and their key characteristics.

4.7.2 *Clinical Indications*

The principal therapeutic uses and structural characteristics of commonly prescribed β-blockers are summarized in Table 4.9, providing insight into their pharmacological diversity and mechanism of action. Beta-blockers are applied across multiple clinical settings:

Hypertension: Used as first-line or adjunctive therapy to reduce blood pressure and cardiovascular risk.

Ischemic Heart Disease: Decrease myocardial oxygen demand, preventing angina and improving post–myocardial infarction survival.

Heart Failure: Certain β_1-selective and α/β-blockers, such as bisoprolol and carvedilol, improve mortality and reverse cardiac remodeling in chronic systolic dysfunction.

Arrhythmias: Control ventricular rate in atrial fibrillation and prevent supraventricular tachycardias via AV nodal blockade.

Other Uses: Propranolol is employed for migraine prophylaxis, essential tremor, performance anxiety, and glaucoma management; beta-blockers also alleviate symptoms of hyperthyroidism.

Table 4.9 Therapeutic classifications and key chemical features of β-blocker drugs

Generic name	Selectivity	Key characteristics
Propranolol	Non-selective	High lipophilicity; crosses blood–brain barrier; used for hypertension, angina, arrhythmias, migraine prophylaxis, essential tremor; also off-label for anxiety and portal hypertension
Nadolol	Non-selective	Long half-life (14–24 h); used in hypertension and angina; low lipophilicity reduces CNS side effects
Timolol	Non-selective	Used for hypertension, prophylaxis of migraine, and reduction of intraocular pressure in glaucoma; intermediate lipophilicity
Sotalol	Non-selective	Class III antiarrhythmic (K^+-channel blockade); used for ventricular and atrial arrhythmias; risk of QT-prolongation and torsades de pointes
Atenolol	β_1-selective (cardioselective)	Low lipophilicity; limited CNS penetration; used for hypertension, angina, and acute MI; once-daily dosing; less risk of bronchospasm but still caution in asthmatics
Metoprolol	β_1-selective	Available as tartrate (immediate-release) and succinate (extended-release); used in hypertension, angina, heart failure, and post-MI; intermediate lipophilicity; may cause fatigue and bradycardia
Bisoprolol	β_1-selective	High β_1 selectivity; used in hypertension and heart failure; minimal effect on β_2-receptors reduces pulmonary side effects
Esmolol	β_1-selective)	Ultra-short-acting (half-life ~9 min); used intravenously for intraoperative tachycardia and emergencies; rapid onset and offset allow tight rate control
Nebivolol	β_1-selective + nitric oxide mediated vasodilation	Highly selective; increases NO availability to induce vasodilation; used in hypertension; favorable metabolic profile; minimal impact on lipid and glucose metabolism
Acebutolol	β_1-selective with intrinsic sympathomimetic activity (ISA)	Partial agonist; reduces resting heart rate less than pure antagonists; used in hypertension and ventricular arrhythmias; moderate lipophilicity
Pindolol	Non-selective with ISA	Partial agonist; less bradycardia at rest; used in hypertension and sometimes in glaucoma (topical); moderate lipophilicity
Labetalol	Non-selective β + α_1-blocker	Combined β- and α-blockade reduces blood pressure without reflex tachycardia; used in hypertension (including pregnancy-induced) and hypertensive emergencies; high lipophilicity
Carvedilol	Non-selective β + α_1-blocker	Antioxidant properties; used in heart failure, hypertension, and post-MI; improves survival in HFrEF; intermediate lipophilicity; may cause weight gain due to α-blockade

(continued)

Table 4.9 (continued)

Generic name	Selectivity	Key characteristics
Betaxolol	β_1-selective	Used orally for hypertension and topically for glaucoma; low risk of bronchospasm; moderate lipophilicity
Celiprolol	β_1-selective with partial β_2-agonism	Unique profile: β_1 antagonism and β_2 agonism; used in hypertension and may improve peripheral blood flow; minimal ISA

4.7.3 Adverse Effects

Common adverse effects include the following:

Cardiovascular: Bradycardia, hypotension, heart block, and reduced exercise tolerance.

Respiratory: Bronchospasm, particularly with nonselective agents, requiring caution in patients with asthma or chronic obstructive pulmonary disease (COPD).

Metabolic: Masked hypoglycemia, altered lipid profiles, and potential weight gain.

Central Nervous System: Fatigue, depression, vivid dreams, or insomnia, especially with lipophilic agents such as propranolol.

Other Effects: Erectile dysfunction, peripheral coldness, and exacerbation of Raynaud's phenomenon.

Beta-blockers are a diverse class of medications that reduce sympathetic stimulation of the heart and vasculature by blocking β-adrenergic receptors, making them essential in managing hypertension, ischemic heart disease, heart failure, and arrhythmias. They are categorized into cardioselective, nonselective, and combined α/β-blockers, with each subclass offering distinct therapeutic benefits and safety considerations. Beyond cardiovascular applications, certain β-blockers are used for migraine prevention, essential tremor, hyperthyroidism symptom control, and anxiety, though careful dosing and patient monitoring are necessary to minimize adverse effects such as bronchospasm, fatigue, and metabolic disturbances.

4.8 Non-steroidal Anti-Inflammatory Drugs (NSAIDS)

NSAIDs are a widely used class of medications that provide analgesic, antipyretic, and anti-inflammatory effects [25]. Their primary mechanism involves inhibition of cyclooxygenase (COX) enzymes, which catalyze the production of prostaglandins responsible for pain, inflammation, and fever. NSAIDs are broadly categorized into nonselective COX-1/COX-2 inhibitors, COX-2 selective agents, and topical NSAIDs, each with distinct efficacy and safety profiles. Nonselective NSAIDs, such

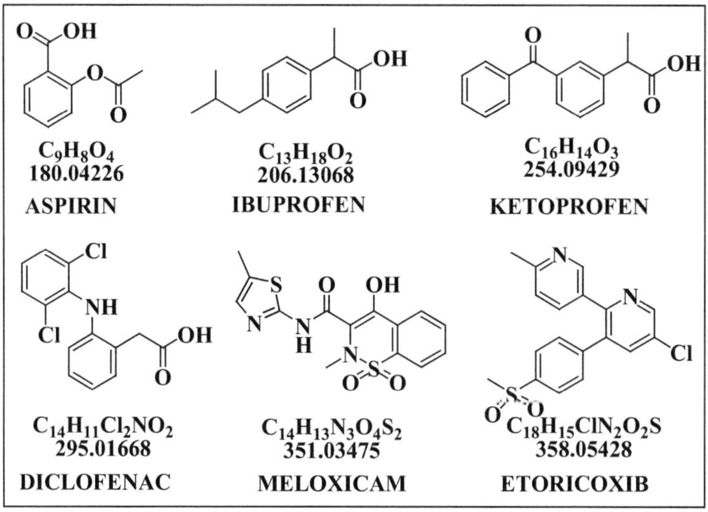

Fig. 4.7 Chemical structure of important NSAIDs

as ibuprofen, naproxen, and diclofenac, inhibit both COX-1 and COX-2, offering potent anti-inflammatory effects but increasing the risk of gastrointestinal toxicity [26, 27]. COX-2 selective inhibitors, including celecoxib, reduce gastrointestinal complications but may elevate cardiovascular risk, particularly in patients with preexisting heart disease. Renal impairment, hypersensitivity reactions, and interactions with other medications such as antihypertensives, anticoagulants, and lithium further influence clinical decision-making. Figure 4.7 depicts the chemical structures of commonly used NSAIDs.

4.8.1 Classification of NSAIDs

Nonselective COX-1/COX-2 Inhibitors: Aspirin irreversibly blocks COX-1, providing antiplatelet benefits at low doses but increasing GI bleeding risk. Ibuprofen and naproxen are reversible inhibitors widely used for pain, fever, and arthritis. Diclofenac and indomethacin are potent inhibitors with notable cardiovascular or CNS-related adverse effects. Ketorolac is primarily used parenterally for short-term severe pain but carries a high risk of gastrointestinal complications.

COX-2 Selective Inhibitors: Celecoxib selectively inhibits COX-2, reducing GI risk while potentially increasing thrombotic events. Other agents such as etoricoxib and rofecoxib have similar profiles, though some have been withdrawn due to safety concerns.

Topical NSAIDs: These formulations, including diclofenac gels and flurbiprofen patches, provide localized effects with minimal systemic exposure.

4.8.2 Mechanism of Action

NSAIDs competitively inhibit COX enzymes, preventing the conversion of arachidonic acid to prostaglandin H_2, a precursor to inflammatory mediators like PGE_2 and thromboxane A_2. COX-1 is constitutively expressed, supporting gastric mucosal protection and platelet function, while COX-2 is inducible at sites of inflammation. COX-1 inhibition accounts for gastrointestinal side effects, whereas selective COX-2 blockade reduces inflammation but may impair endothelial prostacyclin, increasing thrombotic risk.

4.8.3 Clinical Uses

NSAIDs are first-line therapies for a variety of conditions: mild-to-moderate pain such as headache, dysmenorrhea, and musculoskeletal injuries; inflammatory disorders including rheumatoid arthritis, osteoarthritis, and ankylosing spondylitis; and fever reduction via hypothalamic prostaglandin modulation. Specialized indications include acute gout attacks and closure of patent ductus arteriosus in neonates using indomethacin or ibuprofen. The principal NSAID compounds, along with their therapeutic roles, mechanisms of action, and key physicochemical characteristics, are outlined in Table 4.10.

4.8.4 Drug Interactions and Contraindications

NSAIDs can reduce the efficacy of antihypertensives, increase bleeding risk when combined with anticoagulants or SSRIs, and impair the clearance of lithium and methotrexate, increasing toxicity. They are contraindicated in patients with active peptic ulcer disease, severe heart failure, advanced renal disease, or during the third trimester of pregnancy.

4.8.5 Adverse Effects and Risks

Common adverse effects include gastrointestinal complications (dyspepsia, ulceration, bleeding), cardiovascular events (hypertension, myocardial infarction, stroke), renal impairment, and hypersensitivity reactions particularly in aspirin-sensitive

Table 4.10 Summary of major NSAIDs and their pharmacological properties

Generic name	Chemical class	Notable characteristics
Aspirin	Salicylate	Irreversibly inhibits COX-1 and modifies COX-2; antiplatelet at low doses; GI bleeding risk; used for pain, fever, and cardiovascular prophylaxis
Diflunisal	Salicylate	Long-acting aspirin analogue; slower onset, prolonged duration; lower COX-1/COX-2 selectivity; used for chronic arthritis pain, less GI disruption than aspirin
Salsalate	Salicylate	Non-acetylated salicylate; reversible COX inhibition; anti-inflammatory with lower bleeding risk; used in rheumatoid arthritis and osteoarthritis
Ibuprofen	Propionic acid derivative	Reversible nonselective COX inhibition; rapid onset (~1 h), duration ~6–8 h; OTC analgesic/antipyretic; moderate GI and renal risk
Naproxen	Propionic acid derivative	Long half-life (~12–17 h); OTC and prescription formulations; effective in arthritis; lower cardiovascular risk among nonselectives
Ketoprofen	Propionic acid derivative	Potent analgesic/anti-inflammatory; available topically; photosensitivity reactions; moderate GI risk
Flurbiprofen	Propionic acid derivative	High potency; used orally and topically (ophthalmic for post-cataract pain); risk of bronchospasm in aspirin-sensitive patients
Diclofenac	Acetic acid derivative	Very potent nonselective COX inhibitor; used topically and orally; higher cardiovascular risk; effective for arthritis
Indomethacin	Acetic acid derivative	One of the most potent NSAIDs; used in gout flares and patent ductus arteriosus closure; high CNS side effects (headache, depression)
Ketorolac	Acetic acid derivative	Short-term ($\leq$5 days) parenteral analgesic for moderate-severe pain; high GI and renal toxicity; available orally and IV
Piroxicam	Enolic acid (oxicam)	Very long half-life (~50 h); once-daily dosing; effective for Rheumatoid arthritis/Osteoarthritis; higher GI ulceration risk; relatively low platelet effect

(continued)

Table 4.10 (continued)

Generic name	Chemical class	Notable characteristics
Meloxicam	Enolic acid (oxicam)	Preferential COX-2 inhibitor at low doses; lower GI risk than nonselectives; used for arthritis; long half-life (~15 h)
Nabumetone	Prodrug of 6-MNA (acetic acid)	Converted to active COX inhibitor in liver; better GI tolerability; long duration; mild renal effects
Mefenamic acid	Fenamate	Fenamate class; used in dysmenorrhea and mild pain; GI and CNS side effects limit long-term use; reversible COX-1/COX-2 blockade
Meclofenamate	Fenamate	Similar to mefenamic acid; moderate GI toxicity; used for arthritis and menstrual pain; reversible COX inhibition
Celecoxib	COX-2 selective inhibitor	Spares COX-1 (GI protection); increased MI/stroke risk; sulfonamide moiety; used in arthritis and pain management; once- or twice-daily dosing
Etoricoxib	COX-2 selective inhibitor	Highly selective COX-2 blockade; approved in many countries (not US); used for arthritis; lower GI risk; similar cardiovascular profile to celecoxib

asthmatics and hepatotoxicity, notably with diclofenac. Aspirin also causes platelet dysfunction, influencing bleeding risk.

NSAIDs are widely used to reduce pain, inflammation, and fever by inhibiting cyclooxygenase (COX) enzymes, which in turn decreases prostaglandin synthesis. They are classified as nonselective COX-1/COX-2 inhibitors, COX-2 selective agents, or topical formulations, each with distinct therapeutic benefits and safety considerations. While highly effective for conditions ranging from acute musculoskeletal pain to chronic arthritis, their use requires careful monitoring due to potential gastrointestinal, cardiovascular, renal, and hypersensitivity risks.

4.9 Cephalosporins

Cephalosporins are a widely used class of β-lactam antibiotics derived from the fungus *Cephalosporium acremonium*, first isolated in 1948 and developed for clinical use in the 1960s. These agents exert bactericidal activity by inhibiting bacterial cell-wall synthesis through binding to penicillin-binding proteins (PBPs), disrupting peptidoglycan cross-linking and causing osmotic instability that leads to cell lysis [28]. The β-lactam ring of cephalosporins structurally mimics the D-Ala–D-Ala substrate of PBPs, effectively blocking transpeptidation. Compared with penicillins,

Table 4.11 Classification of cephalosporins by generation and spectrum of activity

Generation	Representative agents	Spectrum highlights
First	Cephalexin, Cefazolin, Cefadroxil	Strong Gram-positive activity (Staph, Strep); limited Gram-negative (E. coli, Klebsiella, Proteus)
Second	Cefuroxime, Cefaclor, Cefoxitin, Cefotetan	Retains Gram-positive plus enhanced activity against *Haemophilus influenzae*, *Neisseria*, and anaerobes (cephamycins)
Third	Ceftriaxone, Ceftazidime, Cefotaxime, Cefixime	Expanded Gram-negative including *Pseudomonas* (ceftazidime); good CNS penetration (ceftriaxone)
Fourth	Cefepime	Broad Gram-negative (including *Pseudomonas*), restored Gram-positive; stable versus AmpC β-lactamases
Fifth	Ceftaroline	MRSA and MDR *Streptococcus pneumoniae* activity plus broad Gram-negative; PBP2a affinity

cephalosporins generally show greater resistance to certain β-lactamases, though both antimicrobial spectrum and stability vary across different generations [29, 30].

Cephalosporins are classified into five generations based on antimicrobial activity and β-lactamase stability. Early generations primarily target Gram-positive cocci, while subsequent generations broaden coverage against Gram-negative bacteria. Fifth-generation agents, such as ceftaroline, additionally demonstrate activity against methicillin-resistant *Staphylococcus aureus* (MRSA). This generational classification allows clinicians to tailor therapy to infection severity, site, and likely pathogens, while considering the evolving landscape of bacterial resistance. Table 4.11 provides a summary of the cephalosporin generations and their antimicrobial spectrum.

4.9.1 Clinical Uses

Cephalosporins are employed across a wide range of infections:

Skin and Soft-Tissue Infections: First-generation agents such as cephalexin and cefazolin are first-line for uncomplicated cellulitis and surgical prophylaxis.

Respiratory Tract Infections: Second-generation (cefuroxime) and third-generation (ceftriaxone) agents cover community-acquired pneumonia and otitis media.

Intra-Abdominal Sepsis: Cephamycins such as cefoxitin or cefotetan, often combined with anaerobic coverage, are used for abdominal infections.

Meningitis: Ceftriaxone and cefotaxime achieve therapeutic cerebrospinal fluid concentrations, making them first-choice empiric agents in bacterial meningitis.

Febrile Neutropenia and Nosocomial Infections: Cefepime, with broad-spectrum coverage including Pseudomonas, is a mainstay for high-risk neutropenic fever.

MRSA Skin Infections: Ceftaroline provides an alternative to vancomycin for acute bacterial skin and skin-structure infections.

4.9.2 Resistance Mechanisms

Resistance to cephalosporins arises primarily through β-lactamase production, altered PBPs, and changes in bacterial permeability via efflux pumps or porin mutations. Extended-spectrum β-lactamases (ESBLs) and AmpC enzymes hydrolyze many cephalosporins, while MRSA expresses PBP2a, rendering earlier generations ineffective; fifth-generation cephalosporins overcome this mechanism. Table 4.12 summarizes the main cephalosporin antibiotics, highlighting their generation, antimicrobial spectrum, and clinical applications.

4.9.3 Adverse Effects

Cephalosporins are generally well tolerated, but hypersensitivity reactions ranging from rash to anaphylaxis occur in 1–3% of patients, with 5–10% showing cross-reactivity with penicillins. Gastrointestinal disturbances such as diarrhea and nausea are common, and rare complications include *Clostridioides difficile colitis.* Prolonged use may cause hematologic effects, such as transient neutropenia or thrombocytopenia, and high doses can occasionally lead to renal injury.

Cephalosporins remain versatile and essential antibiotics, offering a structured, generation-based approach for treating infections from mild skin infections to life-threatening sepsis. Rational prescribing, informed by infection type, microbial spectrum, and local resistance patterns, is crucial to maintaining their efficacy in the era of increasing antimicrobial resistance.

4.10 Selective Serotonin Reuptake Inhibitors (SSRIs) and Serotonins-Norepinephrine Reuptake Inhibitors (SNRIs)

Selective serotonin reuptake inhibitors (SSRIs) and serotonin-norepinephrine reuptake inhibitors (SNRIs) are widely regarded as first-line pharmacotherapies for major depressive disorder, various anxiety disorders, and certain chronic pain syndromes [31]. SSRIs function by selectively inhibiting the serotonin transporter (SERT), increasing the availability of serotonin in the synaptic cleft, which contributes to

Table 4.12 Overview of cephalosporin drugs and their attributes

Generic name	Generation	Key spectrum/notes
Cephalexin	First-Gen	Active predominantly against Gram-positive cocci (Staph and Strep spp) and some Gram-negative rods (E. coli, Klebsiella, Proteus)
Cefadroxil	First-Gen	Similar to cephalexin with once-daily dosing; used for skin, soft-tissue, and urinary infections
Cefazolin	First-Gen	Excellent surgical prophylaxis; covers MSSA and streptococci; no MRSA activity
Cefaclor	Second-Gen	Broadens Gram-negative coverage (H. influenzae, E. coli, Klebsiella) but retains Gram-positive activity; used in respiratory and otitis media infections
Cefuroxime	Second-Gen	Enhanced H. influenzae and Neisseria coverage; also used for Lyme disease in children
Cefotetan	Second-Gen	A cephamycin subgroup with anaerobic activity (Bacteroides); used in intra-abdominal infections
Cefoxitin	Second-Gen	Cephamycin subgroup; reliable Bacteroides coverage; used for surgical prophylaxis in colorectal procedures
Ceftriaxone	Third-Gen	Long-acting; excellent CNS penetration; used in meningitis, gonorrhea, and severe systemic infections
Ceftazidime	Third-Gen	Notable Pseudomonas aeruginosa activity; penetrates CSF; used for nosocomial infections
Cefotaxime	Third-Gen	CNS active; often preferred over ceftriaxone in neonates; covers many Gram-negatives
Cefixime	Third-Gen	Oral agent; used for uncomplicated UTI, otitis media, and gonorrhea; less robust Gram-negative spectrum than IV 3rd-gens
Cefepime	Fourth-Gen	Broad Gram-negative (including Pseudomonas) and Gram-positive activity; more stable to β-lactamases
Ceftaroline	Fifth-Gen	Only FDA-approved MRSA-active cephalosporin; covers MRSA, MDR Strep pneumoniae, and broad Gram-negative rods

improved mood and reduced anxiety. SNRIs, in contrast, block both serotonin and norepinephrine transporters (SERT and NET), enhancing synaptic concentrations of both neurotransmitters [32, 33]. This dual action underlies their efficacy not only in mood and anxiety disorders but also in neuropathic pain, with agents like duloxetine and venlafaxine demonstrating significant benefit in diabetic neuropathy and fibromyalgia. The chemical structure of key SSRIs and SNRIs are depicted in Fig. 4.8.

SSRIs are generally preferred for their favorable cardiovascular profile and tolerability, whereas SNRIs can cause mild increases in blood pressure and heart rate. Common adverse effects of SSRIs include gastrointestinal symptoms, headache, sleep disturbances, and sexual dysfunction, while SNRIs often produce nausea, dry mouth, sweating, and hypertension. Both classes carry the risk of serotonin syndrome if combined with other serotonergic agents and require gradual tapering to prevent discontinuation syndromes, which may present with dizziness, irritability,

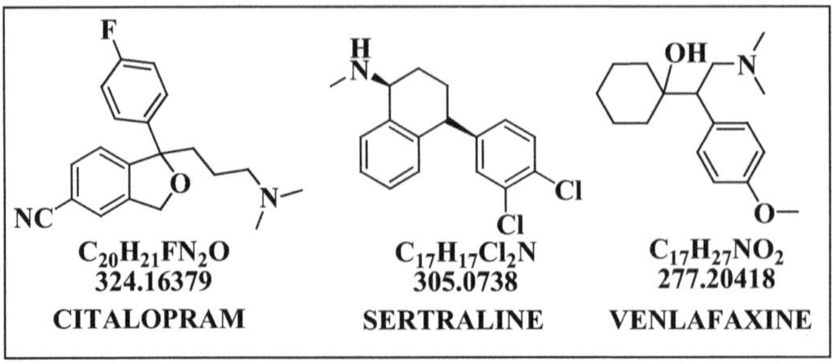

Fig. 4.8 Chemical structure of important SSRIs & SNRIs

or flu-like symptoms, particularly with short-half-life agents such as paroxetine and venlafaxine. Table 4.13 summarizes the pharmacologic profiles, examples, and clinical considerations of commonly prescribed SSRIs and SNRIs.

Table 4.13 Pharmacologic profiles of major SSRIs and SNRIs

Class	Generic name	Affinity	Notes
SSRIs	Fluoxetine	High SERT, minimal NET/DAT	Long half-life (4–6 days) reduces withdrawal; activating, may cause insomnia; many drug interactions
	Sertraline	High SERT	Well-tolerated; slight DAT inhibition; mild stimulation; low drug-interaction risk
	Paroxetine	High SERT	Anticholinergic; more sedating; higher weight-gain risk; discontinuation syndrome common
	Escitalopram (Pure S-enantiomer)	High SERT	Favorable side-effect profile; low drug-interaction; effective for depression/anxiety
SNRIs	Venlafaxine	Moderate SERT/NET	Dose-dependent NET effect; risk of hypertension at high doses; effective in anxiety
	Duloxetine	Balanced SERT/NET	FDA-approved for depression, GAD, diabetic neuropathy, fibromyalgia; watch for hepatotoxicity
	Desvenlafaxine	Active metabolite of venlafaxine	Once-daily dosing; similar profile to venlafaxine; less CYP interactions
	Levomilnacipran	NET-biased SNRI	Greater noradrenergic effect; approved for depression; may increase heart rate/BP

4.10.1 Mechanisms of Action

The therapeutic effects of SSRIs arise from blockade of SERT, preventing presynaptic reuptake of serotonin and elevating extracellular levels to improve mood and anxiety. SNRIs inhibit both SERT and NET, increasing the synaptic availability of serotonin and norepinephrine. The additional norepinephrine modulation enhances analgesic effects in neuropathic pain and augments antidepressant efficacy in more severe presentations. Table 4.14 summarizes the primary pharmacological characteristics, mechanisms of action, and therapeutic uses of commonly prescribed SSRIs and SNRIs.

4.10.2 Clinical Indications

SSRIs and SNRIs are first-line treatments for major depressive disorder across severity levels, with SSRIs often favored for tolerability and SNRIs selected in cases with concomitant pain or more severe depressive symptoms. Both classes are effective for generalized anxiety disorder, panic disorder, and social anxiety disorder. SNRIs, particularly duloxetine and venlafaxine, are also indicated for neuropathic pain, diabetic peripheral neuropathy, and fibromyalgia. Off-label applications include post-traumatic stress disorder, obsessive–compulsive disorder, premenstrual dysphoric disorder, and menopausal vasomotor symptoms.

4.10.3 Safety Considerations

Pharmacogenetic variations in CYP2D6 and CYP2C19 influence SSRI and SNRI metabolism, efficacy, and tolerability, and genotyping may help guide dosing to minimize side effects and nonresponse. Serious adverse events are uncommon but include serotonin syndrome, hyponatremia, and increased suicidal ideation in adolescents and young adults. Overall, SSRIs and SNRIs offer a balance of efficacy and safety, with careful selection, dosing, and monitoring essential to optimize therapeutic outcomes.

SSRIs and SNRIs are first-line treatments for depression, anxiety disorders, and certain chronic pain conditions. SSRIs increase synaptic serotonin, while SNRIs enhance both serotonin and norepinephrine, supporting mood regulation and analgesia. Generally well tolerated, SSRIs favor cardiovascular safety, whereas SNRIs may elevate blood pressure. Adverse effects include gastrointestinal symptoms, sexual dysfunction, and potential serotonin syndrome. Careful dosing, gradual tapering, and pharmacogenetic considerations optimize efficacy and minimize risks.

Table 4.14 SSRIs and SNRIs: Drug profiles and clinical indications

Generic name	Class	Key characteristics
Citalopram	SSRI	Pure serotonin reuptake inhibitor; well-tolerated; risk of QT prolongation at high doses; used for depression and anxiety disorders
Escitalopram	SSRI	S-enantiomer of citalopram; greater potency and selectivity; lower side-effect burden; first-line for depression and GAD
Fluoxetine	SSRI	Long half-life (4–6 days) reduces withdrawal; activating; approved for depression, OCD, bulimia, PMDD; multiple drug interactions via CYP2D6
Fluvoxamine	SSRI	Strong SERT inhibition; indicated for OCD and social anxiety; short half-life; more sedating; significant CYP1A2 inhibition
Paroxetine	SSRI	High anticholinergic and sedative effects; potent CYP2D6 inhibitor; shorter half-life leads to withdrawal symptoms; approved for depression, anxiety, PTSD, PD, PMDD
Sertraline	SSRI	Mild DAT inhibition; well-tolerated; effective in depression, PTSD, PMDD, social anxiety; moderate half-life (~26 h)
Vilazodone	SSRI/ SPARI	SSRI plus 5-HT_1A partial agonist; indicated for MDD; may have lower sexual-dysfunction risk
Vortioxetine	SSRI/ SMRI	Multi-modal: SERT inhibition plus receptor modulation (5-HT_1A agonist, 5-HT_3 antagonist); approved for MDD; may improve cognitive symptoms
Venlafaxine	SNRI	Dose-dependent NET inhibition emerges $\geq$150 mg; risk of hypertension at high doses; used for depression, GAD, panic disorder; significant discontinuation syndrome
Desvenlafaxine	SNRI	Active metabolite of venlafaxine; more predictable pharmacokinetics; approved for MDD; less CYP metabolism; similar efficacy to venlafaxine
Duloxetine	SNRI	Balanced SERT/NET inhibition; approved for MDD, GAD, diabetic neuropathy, fibromyalgia; may elevate liver enzymes
Milnacipran	SNRI	NET-biased SNRI; approved for fibromyalgia; not approved for depression in the US; moderate half-life; fewer CYP interactions
Levomilnacipran	SNRI	Enantiomer of milnacipran; NET-predominant; approved for MDD; may increase HR/BP

4.11 Cannabinoids

Cannabinoids represent a diverse group of compounds, including plant-derived phytocannabinoids, endogenous endocannabinoids, and synthetic analogues, which interact with the body's endocannabinoid system to regulate mood, pain, appetite, immune function, and other physiological processes [34]. The primary psychoactive component of cannabis, Δ^9-tetrahydrocannabinol (Δ^9-THC), acts as a partial agonist at CB_1 and CB_2 receptors, producing euphoria, analgesia, and immunomodulation.

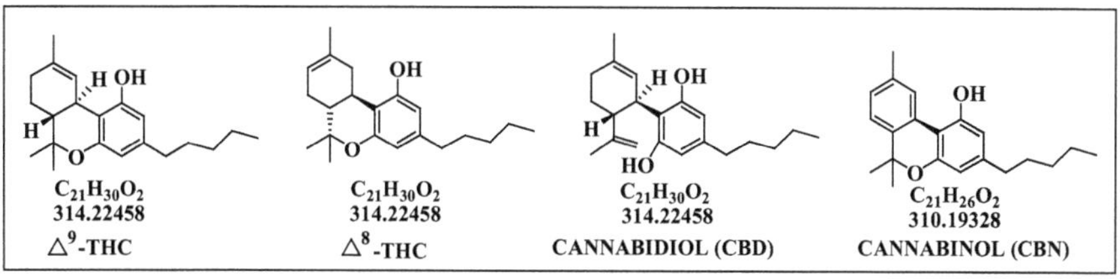

Fig. 4.9 Chemical structures of major cannabinoid compounds

In contrast, cannabidiol (CBD) is non-intoxicating, modulates receptor activity indirectly, and demonstrates anticonvulsant, anti-inflammatory, and anxiolytic effects. Endocannabinoids, such as anandamide (AEA) and 2-arachidonoylglycerol (2-AG), function as neuromodulators in retrograde synaptic signaling, providing fine-tuned control over neurotransmitter release [35, 36]. Clinically approved cannabinoid-based medicines including dronabinol, nabilone, nabiximols, and Epidiolex address conditions such as chemotherapy-induced nausea, AIDS-related anorexia, multiple sclerosis spasticity, and treatment-resistant epilepsy. Ongoing research is exploring minor cannabinoids and the potential "entourage effect," wherein terpenes and less-abundant phytocannabinoids synergize to influence therapeutic outcomes. Figure 4.9 illustrates the chemical structures of important cannabinoids.

4.11.1 *Mechanisms of Action*

Cannabinoids primarily interact with two G-protein-coupled receptors: CB_1, abundant in the central nervous system, and CB_2, expressed mainly on immune cells. Δ^9-THC partially activates both receptors, eliciting psychoactive and immunomodulatory effects, whereas CBD has low direct affinity for CB_1/CB_2 but acts as a negative allosteric modulator and affects other targets such as $5\text{-}HT_1A$ and TRP channels. Endocannabinoids like anandamide and 2-AG are synthesized on demand, act retrogradely to inhibit neurotransmitter release, and are degraded by FAAH and MAGL enzymes, allowing tight regulation of synaptic signaling.

4.11.2 *Phytocannabinoids, Endocannabinoids, and Synthetic Analogues*

Key phytocannabinoids include Δ^9-THC, CBD, and minor cannabinoids such as cannabigerol (CBG), cannabichromene (CBC), cannabinol (CBN), tetrahydrocannabivarin (THCV), and cannabidivarin (CBDV), which are under investigation for neuroprotective, anti-inflammatory, and anticonvulsant effects. Endocannabinoids, including Anandamide (AEA) and 2-Arachidonoylglycerol (2-AG), naturally regulate mood, pain, and appetite through CB_1/CB_2 receptor activity. Synthetic cannabinoids such as dronabinol and nabilone (Δ^9-THC analogues) and nabiximols (THC: CBD oromucosal spray) translate cannabinoid pharmacology into regulated therapeutics for nausea, appetite stimulation, spasticity, and pain.

4.11.3 Therapeutic Applications

Cannabinoids have proven utility in neurological disorders, including refractory epilepsy (CBD/Epidiolex) and multiple sclerosis spasticity (nabiximols). Δ^9-THC and synthetic analogues provide relief for chronic pain, neuropathic conditions, and spasticity, with emerging evidence in fibromyalgia and cancer-related pain. They are also employed as appetite stimulants in HIV/AIDS wasting and cancer cachexia. Potential emerging indications include Parkinson's disease, Tourette's syndrome, sleep disorders, and substance-use disorders, though evidence varies by compound.

4.11.4 The Entourage Effect and Safety

The entourage hypothesis suggests that cannabinoids and other constituents, particularly terpenes like limonene and myrcene, may act synergistically to enhance therapeutic efficacy. Acute effects can include euphoria, anxiety, tachycardia, dry mouth, and dizziness, while chronic or pediatric exposure carries risks of cognitive impairment and dependence, particularly with smoked cannabis. CBD is generally well tolerated but may cause gastrointestinal disturbances or hepatic enzyme elevation. Cannabinoids can also interact with sedatives, antiepileptics, anticoagulants, and CYP450-metabolized drugs. Table 4.15 summarizes key cannabinoids, their pharmacologic characteristics, and primary therapeutic applications.

Phytocannabinoids (Δ^9-THC, CBD) from cannabis contribute to its therapeutic and psychoactive effects, while endogenous cannabinoids (anandamide, 2-AG) regulate synaptic activity and homeostasis via CB_1/CB_2 receptors. Synthetic cannabinoids (dronabinol, nabilone) and botanical extracts (nabiximols) translate these mechanisms into pharmaceutical products used for nausea, appetite stimulation, pain, and spasticity.

Cannabinoids provide a versatile pharmacologic platform comprising phytocannabinoids, endocannabinoids, and synthetic analogues. Rational therapeutic use, informed by mechanism, safety profile, and patient-specific factors, allows for targeted management of nausea, pain, spasticity, appetite loss, and seizure disorders, while ongoing research continues to expand their clinical potential.

Table 4.15 Key cannabinoids: pharmacologic profiles and therapeutic applications

Generic name	Type	Notable characteristics
Δ^9-Tetrahydrocannabinol (Δ^9-THC)	Phytocannabinoid	Principal psychoactive constituent of Cannabis sativa; partial agonist at CB_1/CB_2 receptors; responsible for euphoria, analgesia, and appetite stimulation
Δ^8-Tetrahydrocannabinol (Δ^8-THC)	Phytocannabinoid	Less psychoactive isomer of Δ^9-THC; binds CB_1 with lower affinity; reported anxiolytic and antiemetic effects
Cannabidiol (CBD)	Phytocannabinoid	Non-intoxicating; low affinity for CB_1/CB_2 but modulates endocannabinoid tone; FDA-approved for refractory epilepsy (Dravet, Lennox-Gastaut syndromes)
Cannabinol (CBN)	Phytocannabinoid	Mildly psychoactive; degradation product of THC; potential sedative and antibacterial properties
Cannabigerol (CBG)	Phytocannabinoid	"Stem cell" cannabinoid precursor; non-psychoactive; exhibits anti-inflammatory, analgesic, and neuroprotective effects
Cannabichromene (CBC)	Phytocannabinoid	Non-psychoactive; interacts with TRP channels; potential anti-inflammatory and antinociceptive actions
Tetrahydrocannabivarin (THCV)	Phytocannabinoid	Short-chain THC analogue; CB_1 antagonist/partial agonist depending on dose; studied for appetite suppression and glycemic control
Cannabidivarin (CBDV)	Phytocannabinoid	Propyl analogue of CBD; non-psychoactive; under investigation for epilepsy and neurodevelopmental disorders
Anandamide (AEA)	Endocannabinoid	Endogenous CB_1/CB_2 agonist; involved in mood, pain, and appetite regulation; rapidly degraded by FAAH enzyme
2-Arachidonoylglycerol (2-AG)	Endocannabinoid	Abundant endogenous ligand; full CB_1/CB_2 agonist; key mediator of retrograde synaptic signaling
Dronabinol	Synthetic THC	Pure synthetic Δ^9-THC; FDA-approved for chemotherapy-induced nausea/vomiting and AIDS-related anorexia
Nabilone	Synthetic analogue	THC analogue; FDA-approved for refractory chemotherapy-induced nausea; higher potency and longer duration than dronabinol
Nabiximols	Botanical extract	Oromucosal spray containing ~1:1 THC:CBD; approved in Canada/Europe for multiple sclerosis spasticity and cancer pain; provides balanced cannabinoid ratios

4.12 Antihypertensives (ACE Inhibitors)

Angiotensin-converting enzyme (ACE) inhibitors remain a mainstay in the management of cardiovascular and renal diseases [37]. They function by blocking the conversion of angiotensin I into angiotensin II, a powerful vasoconstrictor, while simultaneously preserving bradykinin, a natural vasodilator. This dual action leads to effective lowering of blood pressure and improved cardiovascular outcomes. Since the introduction of captopril in the 1970s as the first prototype, ACE inhibitors have gained widespread use as first-line agents in hypertension, heart failure, and kidney-related conditions. Their ability to reduce morbidity and mortality has firmly established them as one of the most valuable drug classes in modern clinical practice. Figure 4.10 illustrates the chemical structure of major ACE inhibitors.

The therapeutic benefits of ACE inhibitors extend beyond blood pressure control. They reduce aldosterone secretion, encourage sodium excretion, and improve vascular compliance. Importantly, they play a role in preventing cardiac remodeling and preserving kidney function, especially in patients with diabetic nephropathy [38, 39]. These advantages make ACE inhibitors particularly suitable for individuals at high cardiovascular and renal risk. However, their mechanism also explains certain adverse effects. A persistent dry cough is a well-known side effect, occurring in up to one-third of patients, and in rare cases, angioedema may develop, requiring discontinuation of therapy and consideration of alternative agents such as angiotensin receptor blockers (ARBs).

4.12.1 Mechanism of Action

ACE inhibitors competitively bind to the enzyme ACE, preventing the formation of angiotensin II. As a result, systemic vascular resistance and blood volume decrease, leading to reduced blood pressure. At the same time, the drug preserves bradykinin, enhancing vasodilation through nitric oxide and prostaglandin pathways.

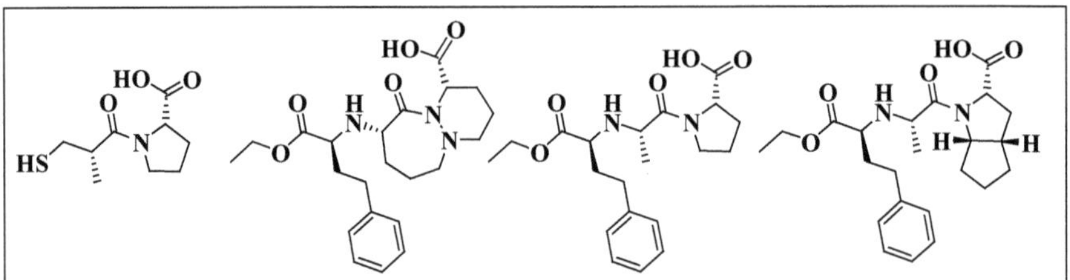

$C_9H_{15}NO_3S$
217.07726
CAPTOPRIL

$C_{22}H_{31}N_3O_5$
417.22637
CILAZAPRIL

$C_{20}H_{28}N_2O_5$
376.19982
ENALAPRIL

$C_{23}H_{32}N_2O_5$
416.23112
RAMIPRIL

Fig. 4.10 Chemical structure of important ACE inhibitors

4.12.2 Clinical Indications

These drugs are widely prescribed for hypertension, post-myocardial infarction care, and heart failure, where they reduce mortality and prevent structural heart changes. They are also key in slowing the progression of diabetic nephropathy and are sometimes used in secondary stroke prevention, Raynaud's phenomenon, and left ventricular dysfunction.

4.12.3 Adverse Effects and Monitoring

Common issues include cough, hypotension, dizziness, and hyperkalemia, while rare but serious cases of angioedema require urgent medical attention. Kidney function and serum potassium should be monitored closely, particularly in patients with existing renal compromise.

4.12.4 Drug Interactions and Contraindications

Caution is necessary when combining ACE inhibitors with NSAIDs, as this may reduce efficacy and impair renal function. They are contraindicated in pregnancy, bilateral renal artery stenosis, and in patients with a prior history of angioedema. Table 4.16 provides a summary of the key characteristics of ACE inhibitors, highlighting their clinical relevance and therapeutic applications.

ACE inhibitors are cornerstone antihypertensives that lower blood pressure, protect the heart and kidneys, and reduce cardiovascular risk. They act by blocking angiotensin II formation and enhancing bradykinin, though side effects include dry cough, hyperkalemia, and rare angioedema.

4.13 Summary

In this chapter, the prohibited framework has been presented as a mainstay of global sports governance, balancing athlete safety, fairness, and the integrity of competition. The review of historical developments, from the early IOC controls to the establishment of WADA and its annual Prohibited List, demonstrates how anti-doping regulation has evolved in response to both scientific progress and recurring challenges in sport.

The categorization of prohibited substances and methods, as outlined in this chapter, reveals the multifaceted nature of doping from anabolic agents that stimulate

Table 4.16 Major ACE inhibitors and their key features

Generic name	Prodrug?	Notable characteristics
Captopril	No	First ACE inhibitor; contains a sulfhydryl moiety; short half-life (2 h); risk of taste disturbance and neutropenia
Enalapril	Yes	Prodrug of enalaprilat; longer half-life (11 h); enalaprilat for rapid BP control
Enalaprilat	No	Active form of enalapril; used in hypertensive emergencies
Lisinopril	No	Long half-life (12 h) allows once-daily dosing; minimal metabolism; renally excreted
Benazepril	Yes	Metabolized to benazeprilat; intermediate half-life (~10 h); often combined with thiazide diuretic
Fosinopril	Yes	Phosphonate-containing prodrug; useful in renal impairment
Moexipril	Yes	Converted to moexiprilat; lipophilic; may improve adherence
Perindopril	Yes	Long-acting; improves outcomes post-MI; available in fixed-dose combos with diuretics
Quinapril	Yes	Converted to quinaprilat; high tissue penetration; used in heart failure and hypertension
Ramipril	Yes	Evidence for reduction in post-MI mortality; long half-life; may improve endothelial function
Trandolapril	Yes	Highly lipophilic; beneficial in heart failure management
Cilazapril	Yes	Long duration; often used in diabetic nephropathy; favorable tolerability

muscle growth to stimulants that enhance alertness, opioids that mask pain, and gene-editing technologies that push ethical boundaries. While the intention of prohibition is clear protecting athletes and preserving trust in sport the constant emergence of new substances and techniques illustrates the dynamic tension between innovation and regulation.

This chapter also emphasized the health and ethical rationale behind prohibitions. Many substances carry severe long-term consequences, undermining athlete well-being in pursuit of short-term gains. Moreover, doping diminishes the essence of sport by transforming contests of skill and dedication into demonstrations of pharmacological advantage. The inclusion of therapeutic use exemptions reflects the complexity of managing genuine medical needs without compromising fairness, while global scandals highlight the need for transparency, accountability, and robust monitoring.

Ultimately, this chapter concludes that the prohibited framework is not static but an evolving system requiring vigilance, collaboration, and adaptation. The spirit of sport must remain the guiding principle, ensuring that athletic achievement is measured by discipline, training, and natural ability rather than artificial enhancement. As doping practices become more sophisticated, the role of regulation becomes even more crucial not merely as enforcement but as a global commitment to safeguarding the authenticity of human competition.

References

1. Lauritzen F, Solheim A. The purpose and effectiveness of doping testing in sport. Front Sports Act Living. 2024;6:1386539. https://doi.org/10.3389/fspor.2024.1386539.
2. Reardon CL, Creado S. Drug abuse in athletes. Subst Abuse Rehabil. 2014;5:95–105. https://doi.org/10.2147/SAR.S53784.
3. Davoren AK, Rulison K, Milroy J, et al. Doping prevalence among U.S. elite athletes subject to drug testing under the World Anti-Doping Code. Sports Med Open. 2024;10:57. https://doi.org/10.1186/s40798-024-00721-9.
4. Thevis M, Kuuranne T, Geyer H. Annual banned-substance review 17th edition—analytical approaches in human sports drug testing 2023/2024. Drug Test Anal. 2025;17:1417–42. https://doi.org/10.1002/dta.3835.
5. WADA/International Testing Authority-The Prohibited List (official site). https://www.wada-ama.org/en/prohibited-list. Accessed 11 June 2025.
6. Heuberger JAAC, Cohen AF. Review of WADA prohibited substances: limited evidence for performance-enhancing effects. Sports Med. 2019;49:525–39. https://doi.org/10.1007/s40279-018-1014-1.
7. Oleksak P, Nepovimova E, Valko M, et al. Comprehensive analysis of prohibited substances and methods in sports: unveiling trends, pharmacokinetics, and WADA evolution. Environ Toxicol Pharmacol. 2024;108: 104447. https://doi.org/10.1016/j.etap.2024.104447.
8. Lu Y, Yan J, Ou G, et al. A review of recent progress in drug doping and gene doping control analysis. Molecules. 2023;28:5483. https://doi.org/10.3390/molecules28145483.
9. Lux D, Vinther AS. Anti-doping research: what is left to do? Drugs Educ Prev Policy. 2023;31:391–5. https://doi.org/10.1080/09687637.2023.2260551.
10. Finley EP, Garcia A, Rosen K, et al. Evaluating the impact of prescription drug monitoring program implementation: a scoping review. BMC Health Serv Res. 2017;17:420. https://doi.org/10.1186/s12913-017-2354-5.
11. U.S. Anti-Doping Agency (USADA). Effects of performance-enhancing drugs. https://www.usada.org/athletes/substances/effects-of-performance-enhancing-drugs/. Accessed 11 June 2025.
12. Leucht S, Priller J, Davis JM. Antipsychotic drugs: a concise review of history, classification, indications, mechanism, efficacy, side effects, dosing, and clinical application. Am J Psychiatry. 2024;181:865–78. https://doi.org/10.1176/appi.ajp.20240738.
13. de Bartolomeis A, Barone A, Begni V, et al. Present and future antipsychotic drugs: a systematic review of the putative mechanisms of action for efficacy and a critical appraisal under a translational perspective. Pharmacol Res. 2022;176: 106078. https://doi.org/10.1016/j.phrs.2022.106078.
14. Alsfouk BA, Aljanadi QM, Almutairi MM. Evaluation of drug interactions in outpatients taking antipsychotic medications. Front Pharmacol. 2025;16:1590167. https://doi.org/10.3389/fphar.2025.1590167.
15. Docherty JR, Alsufyani HA. Pharmacology of drugs used as stimulants. J Clin Pharmacol. 2021;61:S53–69. https://doi.org/10.1002/jcph.1918.
16. Abdulrahim D, Bowden-Jones O. Amphetamine-type stimulants: an overview. In: Textbook of Clinical Management of Club Drugs and Novel Psychoactive Substances: NEPTUNE Clinical Guidance. Cambridge University Press, Cambridge, pp. 146–164 (2022).
17. Avois L, Robinson N, Saudan C, et al. Central nervous system stimulants and sport practice. Br J Sports Med. 2006;40(Suppl 1):i16–20. https://doi.org/10.1136/bjsm.2006.027557.
18. Richards GC, Sitkowski K, Heneghan C, et al. The Oxford catalogue of opioids: a systematic synthesis of opioid drug names and their pharmacology. Br J Clin Pharmacol. 2021;87:3790–812. https://doi.org/10.1111/bcp.14786.
19. Shafi A, Berry AJ, Sumnall H, et al. Synthetic opioids: a review and clinical update. Ther Adv Psychopharmacol. 2022;12:20451253221139616. https://doi.org/10.1177/20451253221139616.

20. Wick JY. The history of benzodiazepines. Consult Pharm. 2013;28:538–48. https://doi.org/10.4140/TCP.n.2013.538.
21. Griffin CE 3rd, Kaye AM, Bueno FR, Kaye AD. Benzodiazepine pharmacology and central nervous system-mediated effects. Ochsner J. 2013;13:214–23.
22. Frishman WH. A historical perspective on the development of β-adrenergic blockers. J Clin Hypertens (Greenwich). 2007;9(Suppl 4):19–27. https://doi.org/10.1111/j.1524-6175.2007.06633.x.
23. Oliver E, Mayor F Jr, D'Ocon P. Beta-blockers: historical perspective and mechanisms of action. Rev Esp Cardiol (Engl Ed). 2019;72:853–62. https://doi.org/10.1016/j.rec.2019.04.006.
24. Ripley TL, Saseen JJ. β-blockers: a review of their pharmacological and physiological diversity in hypertension. Ann Pharmacother. 2014;48:723–33. https://doi.org/10.1177/1060028013519591.
25. Ozleyen A, Yilmaz YB, Donmez S, et al. Looking at NSAIDs from a historical perspective and their current status in drug repurposing for cancer treatment and prevention. J Cancer Res Clin Oncol. 2023;149:2095–113. https://doi.org/10.1007/s00432-022-04187-8.
26. Rao P, Knaus EE. Evolution of nonsteroidal anti-inflammatory drugs (NSAIDs): cyclooxygenase (COX) inhibition and beyond. J Pharm Pharm Sci. 2008;11:81s–110s. https://doi.org/10.18433/J3T886.
27. Khalil NA, Ahmed EM, Tharwat T, et al. NSAIDs between past and present; a long journey towards an ideal COX-2 inhibitor lead. RSC Adv. 2024;14:30647–61. https://doi.org/10.1039/d4ra04686b.
28. Bui T, Patel P, Preuss CV. Cephalosporins. In: StatPearls [Internet]. StatPearls Publishing, Treasure Island (FL) (2025). https://www.ncbi.nlm.nih.gov/sites/books/NBK551517/. Accessed 11 June 2025.
29. Araten AH, Brooks RS, Choi SDW, et al. Cephalosporin resistance, tolerance, and approaches to improve their activities. J Antibiot. 2024;77:135–46. https://doi.org/10.1038/s41429-023-00687-y.
30. Lin X, Kück U. Cephalosporins as key lead generation beta-lactam antibiotics. Appl Microbiol Biotechnol. 2022;106:8007–20. https://doi.org/10.1007/s00253-022-12272-8.
31. Stahl SM, Grady MM, Moret C, et al. SNRIs: their pharmacology, clinical efficacy, and tolerability in comparison with other classes of antidepressants. CNS Spectr. 2005;10:732–47. https://doi.org/10.1017/s1092852900019726.
32. Chu A, Wadhwa R. Selective serotonin reuptake inhibitors. In: StatPearls [Internet]. StatPearls Publishing, Treasure Island (FL) (2025). https://www.ncbi.nlm.nih.gov/sites/books/NBK554406/. Accessed 11 June 2025.
33. Sansone RA, Sansone LA. Serotonin norepinephrine reuptake inhibitors: a pharmacological comparison. Innov Clin Neurosci. 2014;11:37–42.
34. Schurman LD, Lu D, Kendall DA, et al. Molecular mechanism and cannabinoid pharmacology. Handb Exp Pharmacol. 2020;258:323–53. https://doi.org/10.1007/164_2019_298.
35. Bilbao A, Spanagel R. Medical cannabinoids: a pharmacology-based systematic review and meta-analysis for all relevant medical indications. BMC Med. 2022;20:259. https://doi.org/10.1186/s12916-022-02459-1.
36. Shen SY, Wu C, Yang ZQ, et al. Advances in cannabinoid receptors pharmacology: from receptor structural insights to ligand discovery. Acta Pharmacol Sin. 2025;46:1495–510. https://doi.org/10.1038/s41401-024-01472-9.
37. Ahmad H, Khan H, Haque S, et al. Angiotensin-converting enzyme and hypertension: a systemic analysis of various ACE inhibitors, their side effects, and bioactive peptides as a putative therapy for hypertension. J Renin-Angiotensin-Aldosterone Syst. 2023;7890188. https://doi.org/10.1155/2023/7890188.
38. Thind GS. Angiotensin converting enzyme inhibitors: comparative structure, pharmacokinetics, and pharmacodynamics. Cardiovasc Drugs Ther. 1990;4:199–206. https://doi.org/10.1007/BF01857634.
39. Bakhle YS. How ACE inhibitors transformed the renin-angiotensin system. Br J Pharmacol. 2020;177:2657–65. https://doi.org/10.1111/bph.15045.

Chapter 5
Introduction to Biological Matrices in Forensic Science

Abstract This chapter presents an in-depth exploration of biological matrices in forensic science, highlighting their roles, characteristics, and forensic significance. Conventional matrices such as blood and urine remain the backbone of forensic toxicology due to their ability to provide precise, real-time information on toxic substances and drug use. Blood offers unparalleled insight into physiological impairment, while urine is widely used for its non-invasive collection and extended detection window. Beyond these, alternative matrices including hair, nails, sweat, saliva, breast milk, and meconium broaden the scope of forensic analysis by enabling detection over longer or specialized timeframes. These matrices are particularly valuable when conventional specimens are unavailable, degraded, or compromised. Postmortem and veterinary matrices such as vitreous humor, liver, kidney, and bone extend applications into medico-legal and animal-doping contexts, providing unique stability and retrospective exposure data. Advances in analytical technologies, particularly high-resolution mass spectrometry and micro-sampling methods, have revolutionized sensitivity, speed, and scope across these diverse matrices. Each biological specimen presents specific advantages and challenges in terms of collection, stability, and interpretive accuracy, demanding tailored protocols to ensure reliability. Importantly, their forensic applications extend beyond criminal and postmortem toxicology into anti-doping programs, where matrices like blood, urine, and hair underpin the detection of prohibited substances and maintenance of sporting integrity. As a whole, this chapter underscores the scientific, legal, and ethical significance of biological matrices, emphasizing their critical role in strengthening the evidentiary foundation of forensic investigations and anti-doping controls. Future developments in omics technologies and biosensor platforms promise to enhance interpretive power, expand detection panels, and establish biological matrices as indispensable tools for justice, safety, and fair competition.

Keywords Forensic toxicology · Biological matrices · Alternative specimens · Postmortem toxicology · Anti-doping detection · Hair and nail analysis · Vitreous humor · Mass spectrometry

A. K. Karakka Kal et al., *Advanced Extraction Techniques for Drug and Toxin Detection*, https://doi.org/10.1007/978-3-032-17335-5_5

5.1 Biological Matrices

Biological specimens form the backbone of forensic toxicology, offering a wide range of options from conventional fluids to specialized postmortem tissues. Standard matrices such as blood and urine remain central because they provide reliable and well-established information. Blood reflects real-time concentrations of circulating substances, making it invaluable for assessing acute intoxication or impairment. Urine, on the other hand, is widely used for large-scale screening, as it accumulates both parent compounds and their metabolites, thereby extending the detection window beyond immediate exposure [1–3].

When traditional specimens are unavailable or unsuitable, alternative antemortem samples provide powerful supplementary insights. Oral fluid, for example, captures recent drug use within a narrow timeframe of hours, making it particularly useful for roadside testing or workplace monitoring. Hair and nails, in contrast, incorporate substances over weeks to months, allowing reconstruction of long-term exposure histories and serving as important tools in chronic use investigations. Sweat and breath are also increasingly studied for their potential as non-invasive monitoring matrices [4]. Table 5.1 provides an overview of biological specimen categories commonly employed in forensic toxicology.

In postmortem contexts, tissues and fluids such as vitreous humor, cerebrospinal fluid, bone, and liver play a critical role. Vitreous humor is especially valued for its stability, as it resists decomposition and postmortem redistribution, thereby offering more reliable toxicological interpretation when blood and urine are degraded. Similarly, bone and deep tissues like liver and kidney can preserve chemical evidence in cases of advanced decomposition.

Each matrix carries unique advantages and limitations, with considerations such as ease of collection, stability, and analytical sensitivity guiding their selection. While blood and urine continue to dominate routine practice, the expanding use of hair, vitreous humor, meconium, and other specialized samples is significantly enriching forensic investigations. Advances in analytical instrumentation and sample preparation techniques are further enhancing the reliability, precision, and scope of chemical exposure assessments across these diverse biological matrices.

Table 5.1 Categories of biological specimens in forensic toxicology

Conventional matrices	Alternative matrices	Postmortem and veterinary matrices
Blood · Whole Blood · Plasma · Serum	Hair	Liver, stomach contents, and gastric tissues
	Oral fluid (Saliva)	Vitreous humor
	Sweat	Brain and kidney tissues
	Breast milk	Bone and bone marrow
Urine	Nails	Gastric contents
	Breath	Cerebrospinal Fluid (CSF)
	Meconium	

5.2 Blood: Whole Blood, Plasma, and Serum

Blood remains one of the most critical biological matrices in both forensic science and anti-doping efforts. Its unparalleled ability to provide real-time or recent insights into an individual's physiological state, exposure to toxins, or use of illicit substances makes it indispensable in legal investigations and competitive sports monitoring. Unlike other biological fluids, blood accurately reflects current levels of drugs, alcohol, and other xenobiotics, making it the gold standard in forensic toxicology [2, 4]. Its applications range from death investigations and driving-under-the-influence cases to longitudinal athlete monitoring, although its invasive collection and susceptibility to postmortem redistribution are important limitations to consider [5, 6].

5.2.1 Whole Blood

Whole blood contains both the cellular and plasma components of blood, offering a comprehensive snapshot of drug or toxin exposure at the time of collection [7, 8]. In forensic casework, it plays a central role in evaluating impairment, determining causes of death, and reconstructing pharmacokinetic timelines [9]. In anti-doping settings, whole blood strengthens the Athlete Biological Passport (ABP), enabling direct detection of blood-doping practices and exogenous hormone administration.

Characteristics and Advantages: Whole blood contains plasma with clotting factors, along with red and white blood cells and platelets in their natural ratios. This complete matrix preserves drug partitioning between plasma and cells, providing a more accurate measure for compounds that preferentially accumulate in erythrocytes, such as phencyclidine or specific antimalarials. Although the matrix is complex, robust sample preparation methods such as supported liquid extraction effectively remove proteins and cellular debris before chromatographic analysis, ensuring reliable quantification.

Collection, Handling, and Quality Control: Blood is generally collected in K_2EDTA tubes, gently mixed, and either refrigerated (2–8 °C) or frozen (−20 °C/−80 °C) immediately to prevent degradation and continued cellular metabolism. WADA mandates dual-tube collection under strict chain-of-custody protocols, pre-analytical rest periods, and standardized labeling to ensure sample integrity across accredited laboratories. Additionally, freeze-dried whole-blood quality control samples are widely used to monitor inter-laboratory consistency and validate analytical methods.

Forensic Toxicology Applications: Blood alcohol concentration measurements in whole blood remain the legal standard for driving-under-the-influence investigations, correlating directly with psychomotor impairment. Similarly, postmortem whole-blood levels of opioids, sedatives, and other xenobiotics are essential for determining cause of death, requiring careful interpretation in light of postmortem redistribution effects. Serial whole-blood measurements allow pharmacokinetic modeling to estimate time since ingestion or peak drug effects, aiding forensic reconstruction. Advanced high-resolution LC–MS/MS techniques enable simultaneous detection of hundreds of compounds at sub-ng/mL sensitivity, establishing whole blood as the benchmark for comprehensive toxicology panels. Even in scenarios where liquid blood is unavailable, residues from dried stains or clothing can be reconstituted for analysis.

Anti-doping Applications: The hematological module of the Athlete Biological Passport relies on longitudinal monitoring of hemoglobin, hematocrit, and reticulocyte percentages. Unusual variations can indicate practices such as autologous transfusions or micro-dosing with erythropoiesis-stimulating agents (ESAs). Detection techniques including capillary electrophoresis and flow cytometry can identify homologous blood transfusions through minor antigen mismatches between donor and recipient cells. Whole-blood analysis also enables detection of recombinant EPO and other ESAs through isoelectric focusing, differentiating endogenous from exogenous isoforms.

Emerging Techniques: Dried blood spots (DBS) are gaining traction for remote or out-of-competition testing. Blood preserved on filter paper maintains analyte stability while simplifying transportation. Integration with microfluidic extraction and high-throughput automation allows for reduced sample volumes and faster turnaround times without compromising sensitivity. Additionally, emerging omics based approaches may uncover novel biomarkers of doping or intoxication directly from whole blood, enhancing both forensic and anti-doping capabilities.

5.2.2 *Plasma*

Plasma, the cell-free fraction of anticoagulated blood, offers a highly reproducible matrix for quantifying parent drugs, metabolites, and biomarkers [10, 11]. It is widely used in both clinical and forensic toxicology to reconstruct exposure timelines and determine cause of death. In anti-doping programs, plasma is essential for detecting peptide hormones, ESAs, and steroid profiles.

Characteristics and Advantages: Plasma contains circulating clotting factors and unbound xenobiotics, closely reflecting bioavailability in vivo. Compared to whole blood, it is simpler to process, as protein removal is easier, improving chromatographic performance and reducing matrix effects in LC–MS/MS assays. Anticoagulants like K_2EDTA prevent clotting and proteolytic degradation, stabilizing analytes for reliable analysis.

Sample Collection and Processing: Blood is collected into anticoagulant tubes, gently mixed, and centrifuged to separate plasma. The resulting supernatant is aliquoted into pre-labeled, barcoded vials and stored under chain-of-custody conditions at 2–8 °C short-term or −20 °C/−80 °C for long-term storage. Techniques such as protein precipitation or solid-phase extraction concentrate analytes and remove endogenous interferences before instrumental analysis.

Forensic Toxicology Applications: Plasma enables quantitative profiling of drugs and toxins, facilitating pharmacokinetic modeling to estimate time of ingestion and peak effects. High-resolution LC–MS/MS methods allow simultaneous detection of hundreds of compounds from just 50 μL of plasma with sub-ng/mL sensitivity. Plasma remains particularly useful in postmortem investigations when hemolysis or degradation compromises whole-blood analysis, and can be paired with alternative matrices like vitreous humor to strengthen conclusions. Therapeutic drug monitoring in plasma, such as for lamotrigine, supports forensic interpretations in overdose or suicide cases.

Anti-doping Applications: Plasma is central to peptide hormone and growth factor analysis, where immunoassays and LC–MS/MS quantify endogenous versus exogenous proteins. ESA detection uses a technique that separates proteins by their electrical charge, allowing recombinant forms to be distinguished from natural EPO. Plasma also supports the steroidal module of the Athlete Biological Passport, with serial measurements of testosterone, epitestosterone, and metabolites providing longitudinal monitoring for exogenous androgen use. Emerging LC–MS/MS workflows allow both targeted and untargeted analysis, enabling retrospective interrogation for unknown doping compounds.

Quality Assurance and Future Directions: WADA standards and laboratory technical documents dictate strict plasma collection, storage, and analysis protocols to ensure inter-laboratory consistency. Innovations such as dried plasma spots, metabolomics, and high-throughput "omics" analytics promise minimally invasive sampling, expanded biomarker detection, and broader insight into both forensic and anti-doping investigations.

5.2.3 Serum

Serum is the clot-free fraction of blood obtained by allowing whole blood to clot and subsequently centrifuging to remove cells and fibrinogen [12, 13]. Its standardized, cell-free composition improves analytical accuracy, facilitates biomarker profiling, and supports a broad range of toxicological and anti-doping analyses. Unlike plasma, serum lacks anticoagulants, which can interfere with immunoassays or electrophoretic separations, making it the preferred matrix for hormone and peptide studies.

Forensic Toxicology Applications: Serum enables accurate quantitation of a wide range of xenobiotics, from pharmaceuticals such as antidepressants and antipsychotics to illicit drugs and novel psychoactive substances. Validated LC–MS/MS multi-analyte methods achieve detection limits in the sub- ng/mL range. In post-mortem investigations, serum's stability mitigates the effects of hemolysis or post-mortem redistribution, improving interpretive reliability. Serum also allows metabolite profiling, which supports pharmacokinetic modeling and estimation of time since exposure for legal reconstructions.

Serum is suitable for assessing volatile compounds like alcohols or solvents through headspace GC–MS analysis, correlating well with impairment levels. Heavy metals, including lead and arsenic, can also be measured in serum following chelation or microwave digestion protocols, aiding in environmental or occupational poisoning investigations.

Anti-doping Applications: Serum supports immunoassays and LC–MS/MS quantification of peptide hormones such as human growth hormone (hGH) and insulin-like growth factors, distinguishing between endogenous secretion and exogenous administration. ESA detection uses a method that separates proteins based on their electrical charge, and serum provides clearer results with fewer interferences than plasma. The steroidal module of the Athlete Biological Passport uses serial serum measurements of testosterone, epitestosterone, and downstream metabolites to monitor anabolic steroid use.

Emerging Techniques: Dried serum spots have shown promise for preserving hormone and small-molecule stability on specialized filter cards, enabling minimally invasive sampling and simplified logistics for out-of-competition testing. Advances in automation, micro-extraction techniques, and high-resolution mass spectrometry allow for smaller sample volumes, broader analyte panels, and real-time on-site testing.

As summarized in Table 5.2, whole blood, plasma, and serum each offer distinct advantages and limitations for forensic toxicology and anti-doping analyses. Whole blood preserves both cellular and plasma-bound analytes, plasma provides a reproducible cell-free matrix for quantitation, and serum offers a standardized, clot-free matrix suitable for hormone and peptide profiling.

5.3 Urine

Urine represents one of the most widely utilized biological matrices in both forensic toxicology and anti-doping control. Its popularity stems from several factors: non-invasive collection, relatively high analyte concentrations, and the ability to detect drugs and metabolites over extended periods compared to blood. These properties make urine particularly suitable for workplace drug testing, criminal investigations, and anti-doping efforts [14–17]. Typically, urine can reveal recent drug intake within 1–3 days of consumption, although detection windows may vary depending on the substance. Despite these advantages, urinary concentrations often do not correlate directly with the timing of ingestion, which limits its quantitative interpretability in legal proceedings.

Table 5.2 Comparative overview of blood matrices (whole blood, plasma, and serum) for forensic and anti-doping applications

Matrix	Composition/ Characteristics	Advantages	Collection and handling	Forensic applications	Anti-doping applications
Whole Blood	Plasma with clotting factors + all cellular components (RBCs, WBCs, platelets)	Preserves drug partitioning between plasma and cells; reflects true in vivo exposure	K_2EDTA tubes; gentle inversion; refrigerate 2–8 °C or freeze − 20 °C/−80 °C;	DUI cases, postmortem toxicology, pharmacokinetic modeling, cause-of-death assessment	Athlete Biological Passport (hematological module), detection of blood-doping, recombinant EPO/ESA detection
Plasma	Cell-free liquid fraction of anticoagulated blood containing clotting factors and circulating analytes	Simplified protein removal; better chromatographic performance; stable for pharmacokinetic modeling	Collected in anticoagulant tubes; centrifugation 1,500–2,000 × g; aliquoted and stored 2–8 °C short-term, −20 °C/−80 °C long-term	Quantitative toxicant profiling, postmortem investigations, metabolite detection, therapeutic drug monitoring	Detection of peptide hormones, growth factors, EPO isoforms; steroid profiling in ABP
Serum	Clot-free, cell-free fraction of blood obtained after clotting and centrifugation	No anticoagulants (reduces interference); reproducible composition; ideal for hormones and peptides	Clot-activator tubes; clot 30–60 min; centrifuge; store 2–8 °C short-term, − 20 °C/−80 °C long-term	Quantitative drug/toxin measurement, metabolite profiling, volatile toxins, heavy metals, postmortem reliability	Peptide hormones and growth factors, ESA detection, steroidal module of ABP

Recent advances are shaping the future of urine testing. On-site rapid screening devices are being integrated with confirmatory laboratory-based LC-HRMS to improve timeliness and accuracy. Additionally, urinary metabolomics is emerging as a powerful tool for discovering novel biomarkers of exposure or doping. Researchers are also exploring alternative matrices, such as oral fluid or dried urine spots, which aim to reduce risks of sample adulteration and facilitate testing in remote or out-of-competition settings.

5.3.1 *Characteristics of Urine as a Forensic Matrix*

Urine offers unique characteristics that enhance its utility in both forensic and anti-doping contexts. By concentrating both parent compounds and their metabolites, urine often contains higher analyte levels than blood, increasing the sensitivity of detection. This concentration effect allows forensic toxicologists to identify even low-dose exposures, which may be undetectable in other matrices.

Urine provides extended detection windows for many substances: stimulants may be detectable for several hours, whereas cannabinoids or certain benzodiazepines can persist for weeks or even months. Such extended windows make urine a valuable tool for retrospective exposure assessment.

Another advantage is its non-invasive nature. Urine collection does not require specialized medical equipment and can be performed safely with minimal biohazard risk. This property is particularly important in workplace testing or postmortem investigations, where convenience and safety are paramount. To ensure the validity of samples, urine pH and specific gravity are routinely measured. These parameters help detect dilution or adulteration attempts, maintaining the evidentiary integrity of the sample.

5.3.2 *Sample Collection and Integrity*

Urine collection follows strict protocols to guarantee sample authenticity. Typically, samples are collected in tamper-evident containers, and their volume and specific gravity are recorded immediately. Temperature strips confirm that the sample has been freshly voided, preventing potential manipulation.

Adulteration detection is a critical step in urine analysis. Routine checks include creatinine measurement, pH evaluation, and oxidant tests to identify diluted or chemically tampered samples. These procedures ensure that only authentic, uncontaminated specimens proceed to laboratory analysis.

5.3.3 *Forensic Toxicology Applications*

Urine serves a pivotal role in various forensic contexts.

- In impaired-driving investigations, urine analysis can detect alcohol metabolites and psychoactive substances, supporting correlations with blood alcohol concentrations and psychomotor impairment.
- Postmortem toxicology also benefits from urine analysis. Unlike blood, which may be affected by hemolysis or postmortem redistribution, urine often retains detectable concentrations of drugs and metabolites, providing reliable qualitative evidence of antemortem exposure.
- Urine testing is also central to investigations of drug-facilitated crimes. Short-acting sedatives, "date-rape" drugs, and other rapidly metabolized compounds can be detected with high sensitivity using immunoassays followed by confirmatory GC–MS or LC–MS/MS analysis.

- Workplace and probationary drug-testing programs rely heavily on urine due to established cutoff concentrations, rapid high-throughput immunoassays, and cost-effective confirmatory workflows. The matrix provides a practical balance between convenience, sensitivity, and regulatory compliance.

5.3.4 Anti-doping Detection

Urine is a primary matrix in anti-doping programs, as outlined in WADA's regulations. It is effective for detecting a wide array of prohibited substances, including anabolic steroids, stimulants, diuretics, and masking agents, both in-competition and out-of-competition. Urine collection in anti-doping contexts is highly regulated. Athletes are strictly identified before providing a sample under direct observation. Each collection produces an A and B sample, ensuring that any adverse analytical finding can be confirmed independently.

Modern analytical techniques allow multi-class screening of urine samples. GC–MS/MS and LC-HRMS enable the simultaneous detection of traditional and emerging doping agents, including designer drugs. Urinary measurements of endogenous steroid ratios, such as the testosterone/epitestosterone (T/E) ratio or 5α-androstanediol levels, contribute to the steroidal module of the Athlete Biological Passport. This longitudinal monitoring increases sensitivity to subtle or micro-dosing doping strategies.

5.3.5 Analytical Techniques

Several analytical methodologies are employed in urine testing

- Immunoassays: Rapid initial screening for drug classes, such as opioids or cannabinoids, providing immediate results.
- GC–MS/LC–MS/MS: Gold-standard confirmatory methods for precise identification and quantification.
- Isotope Ratio Mass Spectrometry (IRMS): Differentiates between endogenous and synthetic steroids through carbon isotope analysis.
- High-Resolution Mass Spectrometry (HRMS): Detects low-abundance substances, including novel or designer stimulants, enhancing the breadth of detectable compounds.

These complementary techniques ensure both sensitivity and specificity, enabling reliable interpretation of urinary analytes in legal and anti-doping contexts.

5.3.6 Challenges

Despite its widespread use, urine testing faces several challenges. Adulteration remains a significant concern, with athletes or individuals potentially attempting dilution, substitution with synthetic urine, or chemical tampering. Natural variations in urine pH and specific gravity can affect the apparent concentration of analytes, complicating interpretation. Certain substances, such as ESAs or growth hormones, have very short urinary detection windows, necessitating blood analysis for accurate assessment. Ethical and privacy considerations also arise in observed collections, as direct supervision may discourage cooperation in doping tests.

5.3.7 Future Directions

Ongoing innovations continue to strengthen urine testing. Advances in mass spectrometry, high-resolution instruments, and untargeted metabolomics are expanding the range of detectable substances, uncovering novel biomarkers of exposure, and increasing analytical sensitivity. On-site rapid testing devices, coupled with laboratory confirmation, improve the practicality of field-based assessments. Alternative sampling methods, including dried urine spots, enhance stability, reduce shipping logistics, and mitigate adulteration risks.

Overall, urine remains a backbone of forensic toxicology and anti-doping strategies. Its accessibility, high analyte concentrations, and adaptability to sophisticated analytical workflows make it indispensable. While challenges such as adulteration, variable concentration, and short detection windows persist, continuous technological advancements ensure that urine analysis remains a reliable tool for safeguarding legal, occupational, and competitive integrity.

5.4 Hair

Hair serves as a powerful retrospective biological matrix, capturing chemical exposure over weeks to months through segmental analysis. Unlike blood or urine, which primarily reflect recent ingestion, hair incorporates both parent drugs and their metabolites into the growing hair shaft at an approximate rate of 1–1.5 cm per month. Certain compounds, particularly melanin-binding drugs such as cocaine and opioids, accumulate preferentially in hair, making this matrix especially valuable for investigating chronic substance use. However, factors such as environmental contamination and variability in hair color and texture present unique analytical challenges that must be carefully managed [17–21].

5.4.1 Hair as a Forensic Matrix

Hair provides a non-invasive specimen that retains chemical evidence over extended periods, offering a chronological record of exposure. Chemicals enter the hair follicle via the bloodstream, allowing segmental analysis to approximate the timing of ingestion and differentiate single events from repeated or chronic use. Hair is also relatively resistant to postmortem degradation and microbial activity, maintaining analyte integrity even when blood or urine samples are unavailable or compromised. This durability makes hair particularly useful in forensic contexts such as cold cases, drug-facilitated crimes, and neonatal exposure investigations.

5.4.2 Analytical Methods and Sample Preparation

Standard hair collection typically involves cutting 3–5 cm of hair closest to the scalp, representing approximately three to five months of exposure history. When head hair is unavailable, body hair may serve as a substitute, although differences in growth rates must be accounted for. Prior to analysis, hair undergoes thorough decontamination, often using organic solvents, to remove external contaminants. Following cleaning, analytes are extracted through digestion or chemical methods, including acidic, enzymatic, or alkaline protocols, to release incorporated substances for analysis.

Sensitive detection techniques such as LC–MS/MS and GC–MS are then applied to quantify low-level compounds, with or without derivatization.

- **Segmental Analysis**: Hair is divided into sections to reconstruct historical drug use timelines, allowing identification of patterns such as relapses or chronic administration.
- **Melanin Binding Studies**: Account for variations in drug incorporation due to hair pigmentation, recognizing that darker hair may sequester more of certain substances, like cocaine.

5.4.3 Forensic Applications

Hair analysis plays a pivotal role in uncovering both chronic and episodic exposures. In cases of drug-facilitated crime or child abuse, hair can reveal administration of sedatives such as benzodiazepines or prenatal exposure to substances when combined with meconium analysis in neonates. Chronic environmental or occupational exposure to heavy metals (e.g., lead, arsenic) or organic pollutants can also be monitored through hair, providing evidence of long-term cumulative exposure that may not be detectable in blood or urine.

5.4.4 Anti-doping Detection

In sports, hair analysis increasingly complements blood and urine testing by circumventing the short detection windows typical of these traditional matrices. Hair enables the retrospective detection of anabolic steroids, stimulants, and other prohibited substances over months, offering a historical record of doping practices. Experimental studies using both animal models and human volunteers demonstrate that numerous WADA-listed substances can be reliably identified and temporally localized in hair segments.

Out-of-competition testing particularly benefits from hair analysis, as the matrix is highly resistant to post-collection manipulation and unaffected by rapid detoxification techniques that can compromise blood or urine results. Examples include the following:

- **Anabolic Steroids**: Synthetic testosterone and SARMs remain detectable in hair long after use, supporting longitudinal monitoring of athletes.
- **Stimulants and Narcotics**: Substances such as amphetamines, cocaine, and beta-2 agonists (e.g., clenbuterol) can be detected across multiple months, identifying repeated use even if athletes abstain immediately before competitions.

5.4.5 Advantages Over Other Matrices

Hair offers several distinct advantages.

- **Extended Detection Window**: Drugs and metabolites remain detectable for months, unlike the shorter windows in blood and urine.
- **Non-invasive Collection**: Sampling is simpler and less intrusive than blood collection, especially in uncooperative subjects.
- **Resistance to Tampering**: Hair is difficult to adulterate, making it a reliable matrix for monitoring long-term substance use.
- **Environmental Stability**: Hair resists decomposition and microbial degradation, making it ideal for postmortem or cold-case analyses.

5.4.6 Challenges

Despite its utility, hair analysis has inherent limitations. External contamination from environmental sources can confound results, requiring rigorous decontamination protocols to distinguish between systemic exposure and surface deposition. Variability in hair color and texture affects drug incorporation, potentially biasing results across different ethnic groups. Additionally, drugs require approximately 5–10 days to appear in hair following ingestion, limiting utility for assessing acute intoxication. Ethical and cultural concerns may also arise, as hair collection may conflict with religious or personal practices in certain populations.

5.4.7 *Future Directions*

Advances in ultra-sensitive mass spectrometry, micro-segmental analysis, and "omics" approaches promise to enhance the temporal resolution and sensitivity of hair testing while reducing sample size requirements. Integration of metabolomic and proteomic profiling may reveal novel biomarkers for emerging substances and improve differentiation between systemic incorporation and external contamination. These innovations will further solidify hair analysis as a robust, tamper-resistant matrix in both forensic and anti-doping investigations.

In conclusion, hair bridges critical gaps in forensic toxicology and anti-doping efforts, providing a durable and retrospective record of chemical exposure. While challenges such as contamination and melanin bias remain, standardized collection protocols and advancements in analytical techniques continue to enhance its reliability. Hair analysis allows forensic investigators to document chronic abuse or environmental exposures and enables anti-doping authorities to detect long-term performance-enhancing substance use, ultimately supporting evidence-based, ethical decision-making.

5.5 Oral Fluid (Saliva)

Oral fluid, commonly known as saliva, has recently gained recognition as a valuable alternative biological matrix for drug detection in both forensic and anti-doping contexts [22]. Its appeal lies in non-invasive, on-site collection and its strong correlation with the free, unbound fraction of drugs in blood. Compared to blood or urine, saliva sampling simplifies logistics, reduces biohazard risks, and provides timely information on recent drug intake and potential impairment. The slightly acidic pH of saliva favors the trapping of basic drugs, such as amphetamines, enhancing detectability. However, limited sample volumes and the possibility of contamination from smoked substances poses limitations to this selection [23].

5.5.1 *Characteristics and Collection*

Oral fluid is primarily composed of glandular secretions with a small ultrafiltrate derived from blood, reflecting the free fraction of circulating drugs shortly after ingestion. Collection is straightforward, typically using absorbent pads or sponges (e.g., Salivette, Quantisal) placed in the mouth for one to five minutes. This approach yields 1–2 mL of fluid with minimal training required, making it suitable for field and roadside testing. To maintain sample stability, saliva is often buffered or diluted to regulate pH and ionic strength and stored refrigerated (2–8 °C) for short-term use or frozen (−20 °C) for longer-term storage.

5.5.2 *Forensic Toxicology Applications*

Drug and Alcohol Detection: Saliva reflects recent substance intake, making it ideal for roadside or workplace testing. For example, THC from cannabis is detectable in saliva within hours of use, closely aligning with periods of impairment. Portable immunoassay devices enable law enforcement to rapidly screen for stimulants, opioids, and cocaine during traffic stops.

Drug-Facilitated Crimes: Saliva captures short-acting sedatives such as benzodiazepines or GHB within narrow post-ingestion windows, supporting investigations of sexual assault or incapacitation events. Its non-invasive collection makes it particularly suitable when immediate evidence acquisition is critical.

Workplace and Legal Testing: Employers increasingly use saliva for random drug checks due to its ease of collection, though the shorter detection window (hours to days) limits its utility for historical exposure assessment.

DNA Profiling: Saliva left at crime scenes, for example on cigarette butts or drink containers, provides a valuable source of DNA for suspect identification and linkage to forensic evidence.

5.5.3 *Anti-doping Applications*

In sports, oral fluid offers a supplementary matrix to blood and urine, particularly useful for detecting stimulants, narcotics, and certain hormones. Sensitive assays can measure salivary testosterone and estradiol, though pharmacokinetic differences from blood currently limit routine use in doping controls. Reference thresholds for oral fluid steroid profiles have been proposed but require further validation before integration into the Athlete Biological Passport.

Saliva testing reliably detects substances such as amphetamines, cocaine, and opioids, supporting both in- and out-of-competition monitoring of WADA Prohibited List categories S6 (stimulants) and S7 (narcotics). Emerging research also investigates saliva for monitoring stress hormones or synthetic compounds, though blood and urine remain primary matrices for steroid detection.

5.5.4 Advantages

- **Non-invasive**: Enhances compliance among athletes, drivers, and workplace participants.
- **Real-Time Correlation**: Closely reflects circulating drug concentrations, aiding impairment assessment.
- **Reduced Adulteration Risk**: Direct observation during collection minimizes tampering, unlike urine-based methods.

5.5.5 Challenges

- **Limited Detection Window**: Saliva typically reflects only recent use (hours to 1–2 days), making it unsuitable for monitoring chronic exposure.
- **Contamination Risks**: Food, drink, or oral hygiene products can interfere with results.
- **Substance Variability**: Lipophilic drugs such as THC may exhibit low salivary concentrations, reducing sensitivity.
- **Lack of Standardization**: Global differences in cutoff values and collection protocols necessitate harmonized guidelines.

5.5.6 Future Directions

Oral fluid testing continues to evolve with methodological and technological improvements. Micro-sampling techniques and biosensor platforms are enhancing sensitivity for trace-level analytes. Researchers are exploring saliva's potential for detecting peptide hormones and markers of gene doping. Innovations such as dried oral fluid spot technology simplify storage and transport while maintaining analyte stability. Integration of metabolomics and high-throughput "omics" approaches promises to uncover novel biomarkers, improving differentiation between recent use and impairment, as well as detection of designer or emerging substances.

In summary, oral fluid offers a practical, rapid, and minimally invasive matrix for forensic and anti-doping applications. Despite limitations in detection window and standardization, its ease of collection, reduced biohazard risk, and correlation with blood concentrations make it an indispensable alternative matrix in forensic toxicology.

5.6 Sweat

Sweat, a biological fluid secreted by eccrine and apocrine glands, has gained increasing attention in forensic science and anti-doping research due to its non-invasive collection and ability to reflect systemic exposure to drugs, toxins, and performance-enhancing substances [24, 25]. Traditionally overlooked compared to blood or urine, sweat is now recognized as a promising matrix for both real-time monitoring and retrospective analysis. Collected via adhesive patches or wipes, sweat provides cumulative drug exposure data over several days. Although analyte concentrations are generally lower than in blood or urine, sweat is particularly valuable for monitoring compliance in rehabilitation programs or continuous surveillance in legal and anti-doping contexts [26, 27].

5.6.1 Characteristics and Collection

Sweat is secreted by eccrine glands and contains parent drugs, metabolites, and other biomarkers, which are excreted through the skin and captured by collection devices. Adhesive sweat patches, such as PharmChek, are applied to intact skin and sample continuously over 1–10 days, detecting substances used 24–48 h prior to application up to the moment of removal. Patches are designed with skin-friendly adhesives and tamper-evident features, ensuring reliable wear for extended periods without irritation or sample compromise.

Upon removal, sweat patches undergo solvent extraction or solid-phase extraction to recover analytes from the collector pad. High-sensitivity techniques, including GC–MS and LC–MS/MS, allow quantification of multiple drug classes amphetamines, opioids, cannabinoids, and benzodiazepines in a single analytical run.

5.6.2 Forensic Applications

- **Drug Use Monitoring**: Sweat analysis is effective for detecting short-acting sedatives like benzodiazepines or GHB in drug-facilitated assault investigations. Continuous patch sampling reduces the risk of missing episodic drug use and provides compelling evidence in probation or workplace compliance programs.
- **Long-Term Surveillance**: Sweat patches capture cumulative exposure over days to weeks, making them ideal for monitoring probation, parole, or rehabilitation programs. Metabolites of cocaine, opioids, and cannabis remain detectable, offering evidence of chronic use.
- **Postmortem Toxicology**: In scenarios such as mass-casualty events or severely decomposed remains, sweat residue on clothing or personal items may serve as an

alternative matrix for toxicological screening. Sweat testing also provides insights into environmental toxin exposure, including heavy metals or ethanol, when blood or urine are unavailable.
- **Touch DNA and Trace Evidence**: Sweat deposits on surfaces such as weapons or textiles can contain epithelial cells for DNA profiling, linking individuals to crime scenes.
- **Environmental Exposure**: Sweat analysis detects bioaccumulated toxins, such as lead or mercury, relevant in occupational or poisoning cases.

5.6.3 Anti-doping Applications

Sweat patches enable continuous monitoring in rehabilitation and drug-diversion programs, offering detection windows corresponding to patch wear rather than single collection points. Although not yet mandated by WADA, sweat testing complements urine and blood analyses, detecting substances like opioids, amphetamines, and cannabis, and helping deter micro-dosing strategies. Emerging research explores sweat for performance-enhancing agents, including anabolic steroids and peptide hormones, with high-resolution mass spectrometry improving detection sensitivity. Applications include the following:

- **Stimulants and Narcotics**: Amphetamines, cocaine, and opioids are excreted in sweat, providing an alternative to urine testing.
- **Anabolic Agents**: Lipophilic steroids, such as testosterone, diffuse into sweat, providing an alternate access for scrutiny.
- **Wearable Sensors**: Epidermal patches can monitor athletes in real time, flagging substances such as beta-2 agonists or diuretics during training.

5.6.4 Advantages

- **Non-invasive**: Sweat collection avoids privacy concerns associated with blood or urine sampling.
- **Extended Detection Window**: Patches provide cumulative data over days, in contrast to blood's short-term snapshot.
- **Resistance to Tampering**: Continuous collection reduces the risk of sample adulteration.
- **Versatility**: Compatible with both laboratory analysis and portable biosensor technologies.

5.6.5 Challenges

- **Low Analyte Concentrations**: Many substances, including growth hormones, occur at trace levels, requiring ultra-sensitive instruments.
- **Variable Secretion Rates**: Sweat composition varies with hydration, physical activity, and individual physiology, complicating interpretation.
- **Contamination Risks**: Environmental exposure to drugs on the skin may produce false positives.
- **Lack of Standardization**: Universal cutoff values and collection protocols are still lacking for forensic and anti-doping applications.

Sweat offers a continuous, non-invasive matrix for monitoring substance use in forensic and anti-doping settings. Despite challenges like low analyte levels, advancements in wearable sensors, microfluidics, and high-resolution analytics enhance its reliability, complementing traditional matrices and enabling broader applications in substance detection, health monitoring, and personalized medicine.

5.7 Meconium

Meconium, the first stool passed by a newborn, is a distinctive biological matrix formed during fetal development. It consists of materials ingested in utero, including amniotic fluid, epithelial cells, mucus, lipids, and bile acids. Its composition makes it an invaluable tool for forensic toxicology, particularly for detecting prenatal drug exposure. While meconium's relevance to anti-doping remains largely theoretical, its principles inform long-term biomarker monitoring strategies such as hair or nail analysis. Formed from around 12–16 weeks of gestation and accumulating through birth, meconium captures a wide range of substances from the second and third trimesters, providing a detection window far longer than maternal urine or blood [28, 29]. Lipophilic drugs and metabolites readily incorporate into meconium, creating a concentrated repository for xenobiotic screening.

5.7.1 Formation and Detection Window

Meconium begins forming in the fetal intestine during gestational weeks 12–16, with the bulk deposited after week 38. Maternal substances consumed during this period, including medications, illicit drugs, and alcohol, are transferred to the fetus through amniotic fluid. Because the fetus continually swallows amniotic fluid, drugs and metabolites accumulate in meconium at concentrations often higher than in urine, creating a retrospective exposure record spanning approximately 20 weeks. This long detection window enables identification of both chronic and episodic maternal substance use.

5.7.2 Sample Collection and Preparation

Standard collection involves retrieving the first three stools passed by the neonate, usually within 48 h of birth, using clean, solvent-washed containers to prevent contamination. Samples are homogenized to ensure uniformity, then undergo extraction using solvent precipitation or solid-phase extraction to isolate parent drugs and metabolites. Rigorous decontamination steps, such as washing with organic solvents, remove external contaminants, ensuring that positive findings accurately reflect in utero incorporation rather than postnatal exposure.

5.7.3 Forensic and Clinical Applications

Meconium provides a long-term biomarker of prenatal exposure, capturing maternal substance use across the second and third trimesters. Unlike maternal urine or blood, which only reflect recent intake, meconium offers an extended detection window. Its forensic applications include documenting maternal drug abuse in child welfare investigations, custody disputes, or sudden infant death assessments. Clinically, meconium testing identifies neonates at risk for neonatal abstinence syndrome, guiding early interventions.

Key applications include the following:

- **Child Welfare**: Detecting illicit drug exposure (opioids, cocaine, methamphetamines) to initiate protective measures.
- **Legal Proceedings**: Providing evidence in custody cases or charges of child endangerment.
- **Public Health Monitoring**: Tracking maternal substance use trends within populations.

5.7.4 Anti-doping Considerations

Although primarily used in neonatal toxicology, meconium principles inform anti-doping strategies, particularly for long-term biomarker monitoring. Cord-tissue or similar fetal matrices reflect cumulative exposure, akin to meconium's 20-week record. While maternal doping during pregnancy is rare, methodological approaches developed for meconium, such as detecting long-term metabolite accumulation, may inspire future techniques for monitoring chronic performance-enhancing drug use.

5.7.5 Advantages and Limitations

Advantages

- Extended detection window from mid-gestation through birth.
- Non-invasive collection post-birth via stool.
- High analyte concentrations improve sensitivity for lipophilic substances.

Limitations

- Sample heterogeneity requires thorough homogenization.
- Processing is labor-intensive due to complex decontamination and extraction steps.
- Interpretation can be confounded by substances administered during labor or neonatal care.

Meconium analysis offers a non-invasive, long-term record of prenatal drug exposure, supporting forensic, clinical, and public health applications. Advances in mass spectrometry and standardized protocols enhance sensitivity and reliability while safeguarding neonatal health and maternal rights.

5.8 Breast Milk

Breast milk, the primary source of nutrition for infants, is a unique biological matrix with significant potential in forensic and toxicological investigations. Produced by lactating individuals, it reflects not only maternal health but also systemic exposure to drugs, toxins, and environmental pollutants. Although breast milk is not commonly included in standard forensic or anti-doping protocols, its analysis provides valuable insights in specialized contexts. By capturing both acute and chronic maternal exposures, breast milk testing informs risk assessment for infants, aids forensic casework, and opens possibilities for novel anti-doping research [30–32].

Breast milk is particularly effective at accumulating lipophilic drugs and metabolites. This occurs through passive diffusion and active transport from maternal plasma, creating a detection window that spans several hours to days before sampling. Substances such as cannabinoids, cocaine, ethanol, opiates, and nicotine have been reliably identified in breast milk, using both immunoassay screening methods and confirmatory chromatographic techniques. Case studies and population surveys highlight its value in assessing maternal substance use and infant exposure, while its utility in sports toxicology remains exploratory.

Recent advances in analytical chemistry are further enhancing breast milk testing. Micro-extraction strategies, direct-analysis approaches such as DART-MS, and high-resolution LC–MS platforms have improved sensitivity and throughput, even with very small sample volumes. These technologies make breast milk a more practical tool for surveillance and casework. At the same time, ethical safeguards regarding informed consent, maternal privacy, and infant health remain central to its responsible use. Establishing standardized collection procedures, validated cut-off values, and transparent reporting practices will be essential for its broader application.

5.8.1 Mechanisms of Drug Transfer into Breast Milk

Drugs typically enter breast milk through passive diffusion across mammary epithelial cells. The degree of transfer depends on several factors, including lipid solubility, molecular size, plasma protein binding, and pKa. Compounds that are highly lipophilic and of low molecular weight, such as Δ^9-THC, tend to concentrate in milk fat, extending detection beyond conventional blood or urine matrices. This makes breast milk particularly useful for monitoring substances with longer half-lives.

5.8.2 Forensic Science Applications

Infant Exposure Investigations: Breast milk testing can reveal exposure to sedatives, opioids, cocaine, or alcohol, which may contribute to infant harm. This evidence is crucial in cases of neglect, poisoning, or custody disputes.

Maternal Death Investigations: Postmortem analysis of breast milk can identify drugs or toxins that contributed to maternal death, especially when other matrices are degraded. Concentration data also help estimate the infant's likely exposure and guide clinical responses to neonatal withdrawal or toxicity.

Environmental Monitoring: Breast milk is an effective biomarker for pollutants such as pesticides and polychlorinated biphenyls, making it valuable in environmental toxicology and public health surveillance.

5.8.3 Doping Detection Considerations

Although not yet adopted by the WADA, breast milk represents a promising non-traditional matrix for long-term monitoring. Substances such as anabolic steroids, stimulants, or peptide hormones could theoretically be tracked, extending the Athlete Biological Passport framework into lactational phases.

Maternal Athletes: Breast milk could inadvertently reveal the use of performance-enhancing drugs (PEDs) by postpartum athletes, raising concerns about infant exposure.

Therapeutic Use Exemptions (TUEs): Medications permitted for athletes may still need to be assessed for their transfer into milk to ensure infant safety.

5.8.4 Advantages

- Non-invasive collection compared to blood sampling.
- Direct insights into infant exposure to maternal substances.
- Extended detection window for lipophilic compounds such as cannabis.

5.8.5 Challenges

- Low analyte concentrations that require sensitive instrumentation.
- Variable milk composition, especially fat content, complicates quantification.
- Ethical and privacy concerns, especially around testing without consent.
- Lack of standardized thresholds or global protocols for forensic and anti-doping contexts.

Breast milk analysis is a specialized yet invaluable tool in forensic science and maternal-infant health. It provides unique information on infant exposure to drugs, toxins, and pollutants, while also supporting investigations of maternal fatalities and child welfare cases. Although its role in doping detection is still theoretical, breast milk highlights the importance of protecting nursing infants from harmful exposures. With continuing advances in analytical techniques and stronger ethical frameworks, breast milk has the potential to become a powerful complementary matrix that bridges health, law, and human rights.

5.9 Nails

Nails, composed mainly of keratin, represent an increasingly valuable biological matrix in forensic toxicology and anti-doping investigations. Their slow and consistent growth, coupled with their ability to sequester a wide range of substances, allows them to provide a retrospective record of drug, toxin, or performance-enhancing agent exposure. Unlike blood and urine, which reflect recent use, nails capture longer-term patterns, often spanning several months. Their non-invasive collection, ease of storage, and resistance to environmental adulteration make them a useful complement to more traditional biological samples. Similar to hair, nails provide a long detection window but are generally less prone to external contamination or cosmetic interference. These qualities make nail analysis particularly advantageous in chronic exposure cases where long-term histories must be reconstructed [33, 34].

The development of advanced extraction procedures and highly sensitive mass-spectrometric methods has broadened the scope of analytes detectable in nail clippings. Substances such as opioids, stimulants, anabolic steroids, and even metabolic biomarkers can now be detected at trace levels (ng/mg), strengthening the evidentiary value of this matrix. While nail growth rate variability and differences in drug incorporation remain challenges, emerging guidelines and ongoing standardization efforts are enhancing both the reliability and legal defensibility of nail testing.

5.9.1 *Biological Basis and Detection Window*

Nails are formed at the germinal matrix and grow outward at an average rate of about 3 mm per month. During their formation, compounds are incorporated through blood supply to both the nail matrix and nail bed. This dual pathway ensures that drugs and metabolites are retained within the keratin structure for extended periods. On average, nails provide a detection window of 3–6 months, with the maximum retrospective period corresponding to the full age of the nail plate, typically around half a year. Lipophilic substances, such as cannabinoids or anabolic steroids, integrate particularly well, offering extended timelines of detection. Unlike hair, nail analysis is not influenced by ethnicity or cosmetic treatments, making it a more consistent matrix across populations.

5.9.2 *Collection and Preparation*

- **Sampling**: Nail clippings are collected non-invasively, usually from all ten digits, to maximize historical coverage. Clippings close to the nail bed provide the most recent exposure profile.
- **Decontamination**: To minimize environmental contamination, samples undergo sequential washes, ensuring that only incorporated compounds are measured.
- **Extraction**: Nail keratin is digested using alkaline hydrolysis, enzymatic treatment, or microwave-assisted extraction. Analytes are then isolated through solid-phase or liquid–liquid extraction before analysis.

For initial screening, immunoassay techniques such as ELISA provide fast, semi-quantitative testing for major drug groups like opioids, amphetamines, and benzodiazepines. For definitive analysis, high-resolution methods including GC–MS/MS and LC–MS/MS deliver precise quantitative confirmation, often achieving detection limits in the low picogram range, with isotope-labeled internal standards used to ensure accuracy and consistency.

5.9.3 *Forensic Science Applications*

Chronic Exposure Assessment

- Toxic metals such as arsenic, lead, or mercury accumulate in nails, allowing investigators to document long-term poisoning in both criminal and environmental cases.
- Segmental analysis of clippings provides timelines of drug use, revealing patterns of cocaine, opioids, or prescription medication intake.

Postmortem Investigations

- Nails persist long after soft tissue decomposition, making them useful in skeletal remains.
- Growth patterns can help estimate the timing of substance use before death, offering time-stamped toxicological evidence.

Criminal and Civil Cases

- Nail testing can influence custody disputes by establishing chronic substance use in caregivers.
- Occupational exposure to hazardous chemicals can also be verified through nail analysis, supporting workplace safety claims.

5.9.4 Doping Detection Applications

Nail testing has potential in anti-doping programs, particularly where extended monitoring is required.

- **Steroids and Designer Drugs**: Lipophilic anabolic steroids such as testosterone integrate into nails, remaining detectable for months after use.
- **Alcohol Biomarkers**: Ethyl glucuronide and fatty acid ethyl esters incorporated into nails act as indicators of chronic alcohol misuse, similar to hair testing.
- **Stimulants and Diuretics**: Extended detection windows make nails suitable for uncovering micro-dosing or drug cycling strategies intended to evade traditional blood and urine tests.

5.9.5 Advantages

- Long detection window, capturing substance use over several months.
- Stable matrix resistant to degradation, enabling long-term storage.
- Non-invasive and discreet collection with fewer privacy concerns than urine sampling.

5.9.6 Limitations

- Variation in nail growth among individuals complicates precise timeline reconstruction.
- Surface contamination from environmental sources requires rigorous decontamination.
- Analytes are often present at trace levels, demanding advanced instrumentation.
- Cultural or ethical objections to nail collection may occasionally limit feasibility.

Nail analysis provides a powerful, non-invasive method for assessing long-term substance exposure, bridging forensic toxicology and anti-doping science. Its ability to retain drugs, toxins, and biomarkers for months makes it especially valuable in chronic exposure investigations, postmortem analysis, and extended doping surveillance. Although challenges such as contamination risks, growth variability, and low analyte concentrations remain, advances in extraction methods and analytical sensitivity continue to improve its applicability. Future directions will likely include the development of standardized international guidelines, incorporation of emerging biomarkers, and miniaturized on-site testing technologies. As research progresses, nail testing is poised to become a essential part in evidence-based toxicology, ensuring both justice in forensic contexts and fairness in competitive sports.

5.10 Breath

Breath analysis has emerged as a transformative tool in forensic science and anti-doping research, providing a rapid, non-invasive, and ethically sound alternative to conventional matrices such as blood and urine [35, 36]. Traditionally associated with alcohol detection, it has expanded significantly in scope with the advent of advanced analytical technologies. Today, breath is recognized as a versatile biological matrix capable of detecting a wide spectrum of drugs, toxins, and performance-enhancing substances, making it a valuable complement to existing testing strategies [37, 38].

5.10.1 Historical Context and Evolution

The forensic application of breath testing dates back to the 1920s, when early breath-alcohol analyzers were first introduced for traffic law enforcement. Over time, these devices became the gold standard in assessing blood alcohol concentration during impaired driving cases. In the last two decades, research has moved beyond alcohol to contain a broader range of analytes. This shift has been facilitated by improvements in sensitivity and specificity, enabling detection of volatile drugs of abuse and even some performance-enhancing agents.

A variety of methods are employed in breath testing, each with its own strengths:

- **GC–MS**: Provides high sensitivity for detecting volatile organic compounds (VOCs), including drugs and metabolites.
- **PTR-MS**: Allows real-time detection of trace compounds with minimal preparation.
- **Sensor-Based Devices**: Portable breathalyzers and "electronic noses" use nanosensor arrays to screen substances on-site, offering practical solutions for roadside or field investigations.
- **Exhaled Breath Condensate (EBC)**: By cooling exhaled air, this method captures non-volatile components such as proteins and lipids, broadening the range of detectable biomarkers.

5.10.2 Forensic Science Applications

Breath remains indispensable in alcohol testing, providing immediate results that correlate with impairment levels. Its utility now extends to detecting substances like cannabis (THC), opioids, and amphetamines shortly after consumption. For example, THC has been measured in exhaled air up to four hours after use, which is critical for roadside impairment assessment. Breath testing has also proven valuable in identifying exposure to volatile toxins such as carbon monoxide or cyanide, aiding emergency interventions.

Although less common, breath analysis has been applied in postmortem investigations. Exhaled breath condensate from recently deceased individuals may retain VOCs, offering additional evidence in cases involving ethanol or inhalant exposure.

5.10.3 Doping Detection Applications

Proof-of-concept studies have demonstrated the feasibility of breath testing for anti-doping purposes. Such tools reduce athlete burden, providing a discreet and minimally invasive testing option. Breath-based monitoring shows particular promise in detecting cannabinoids and stimulants, both relevant in competitive sports. Although challenges remain in identifying non-volatile compounds such as peptide hormones, ongoing research is exploring breath biomarkers and metabolomic approaches.

5.10.4 Advantages

- **Non-invasive collection**: Eliminates discomfort and privacy issues linked to blood and urine.
- **Rapid results**: Delivers immediate findings, enabling on-the-spot decision-making.

- **Minimal adulteration risk**: Collection under observation reduces tampering opportunities.
- **Versatility**: Applicable in law enforcement, sports, occupational safety, and environmental monitoring.

5.10.5 Challenges

Despite its promise, breath testing faces limitations. Most analytes remain detectable for only a few hours, restricting its value for long-term monitoring. Many drugs occur at trace levels in breath, necessitating ultra-sensitive instruments. Environmental contaminants, such as secondhand smoke or industrial pollutants, can interfere with results. Furthermore, the absence of standardized cutoff values and collection protocols complicates interpretation across forensic and sports settings.

Breath analysis is redefining substance detection by combining practicality with scientific rigor. Although constrained by short detection windows and the need for standardization, it offers unmatched advantages in speed, non-invasiveness, and ethical acceptability. Future directions include integrating metabolomic workflows, refining EBC techniques for non-volatile compounds, and applying artificial intelligence to interpret complex VOC patterns. Collaborative efforts with regulatory bodies such as the WADA will be essential for establishing global standards. Innovations in wearable sensors and continuous monitoring devices may eventually enable real-time oversight in athletic and forensic contexts. Collectively, these advancements underscore breath testing's potential to enhance public safety, safeguard fairness in sport, and promote accountability in diverse environments.

5.11 Liver, Stomach Contents, and Gastric Tissues

Postmortem and veterinary biological matrices such as the liver, stomach contents, and gastric tissues serve as valuable resources for toxicological and forensic investigations [39, 40]. Unlike conventional matrices like blood and urine, these tissues often retain parent drugs and metabolites for longer periods, making them particularly useful when routine samples are unavailable or degraded. Their ability to extend detection windows provides critical insights into exposure history and enhances the reconstruction of substance intake timelines. However, these matrices are inherently complex, with challenges such as postmortem redistribution, lipid-rich environments, and tissue degradation that demand robust analytical protocols. Recent developments in chromatographic and mass spectrometric techniques, combined with improved sample preparation strategies, are helping to overcome these obstacles and strengthen the evidentiary value of findings in both human forensic science and veterinary doping control.

5.11.1 Liver

Due to its central role in metabolism and high blood supply, the liver accumulates significant levels of lipophilic drugs and their metabolites [41]. In many cases, concentrations in liver tissue exceed those detected in blood, providing an advantage when examining bodies in advanced decomposition. State-of-the-art techniques, such as ultra-high-performance liquid chromatography coupled with tandem mass spectrometry (UHPLC-MS/MS), allow precise detection of pharmaceuticals and illicit substances at nanogram-per-gram levels, even in this challenging matrix. Nevertheless, interpretation requires caution, as postmortem redistribution can cause drugs to migrate between tissues and fluids, complicating the estimation of concentrations prior to death.

5.11.2 Stomach Contents

Stomach contents provide a direct reflection of recent oral intake, offering toxicologists evidence about the timing and quantity of ingested substances [42, 43]. Their complex composition, often rich in lipids and partially digested matter, poses analytical difficulties. To address this, specialized clean-up techniques such as solid-phase extraction with lipid-removal cartridges have been validated to improve analyte recovery. Advanced protocols using GC–MS and LC-HRMS have successfully detected opioids, stimulants, and designer drugs in gastric aspirates, even in cases where blood or urine samples were unavailable. Furthermore, gross examination of stomach contents, including intact pill fragments or botanical material, provides complementary physical evidence during autopsy investigations.

5.11.3 Gastric Tissues and Mucosa

In addition to luminal contents, gastric mucosa and adjacent tissues provide unique insights into corrosive injuries or localized drug deposition [44]. Histological examinations have shown tissue necrosis and inflammatory changes following ingestion of caustic agents, reinforcing toxicological evidence from gastric fluid. Since these tissues are susceptible to redistribution of small molecules after death, standardized sampling methods considering depth and location are essential to reduce variability and improve interpretive accuracy.

5.11.4 Implications for Doping Control

Although routine anti-doping testing relies primarily on blood and urine, postmortem and veterinary matrices extend investigative capacity, particularly in retrospective or clandestine cases. For example, anabolic steroids and β_2-agonists may persist in liver tissue for weeks after administration, providing critical proof of illicit use in equine sports or animal competitions. Similarly, gastric contents may retain intact drugs or formulation excipients, revealing both the identity of substances and their approximate timing of administration. These applications highlight the broader utility of such matrices in promoting fairness and integrity in sports and animal welfare.

5.11.5 Challenges and Future Perspectives

Despite their potential as a valuable source of forensic evidences, several limitations remain. Postmortem redistribution complicates quantitative interpretation, while matrix-specific interferences can obscure results. Efforts to harmonize protocols across forensic and veterinary laboratories are ongoing. Innovative approaches such as probe electrospray ionization mass spectrometry (PESI-MS) aim to minimize sample preparation while maintaining high sensitivity. At the same time, metabolomic profiling promises to expand detection capabilities to emerging designer drugs and novel veterinary growth promoters. Establishing consensus guidelines, similar to those of international forensic bodies like TIAFT, will be vital to ensure reproducibility and legal defensibility.

Liver, stomach contents, and gastric tissues remain indispensable in forensic and veterinary toxicology. They provide unique insights into drug exposure, administration routes, and toxic mechanisms that often cannot be obtained from blood or urine alone. By embracing advanced analytical technologies and standardized sampling methods, these matrices will continue to play a pivotal role in uncovering hidden toxins, resolving complex forensic cases, and safeguarding the integrity of competitive animal sports and food safety.

5.12 Vitreous Humor

Vitreous humor (VH) has established itself as an indispensable matrix in postmortem toxicology and is increasingly being explored for retrospective doping investigations. Its unique characteristics, chemical stability, and transport mechanisms that differ significantly from blood and urine make it particularly valuable in cases where conventional samples are unavailable or degraded [45, 46]. Located within the posterior chamber of the eye, VH is well shielded from putrefaction and microbial invasion, making it less prone to contamination and postmortem redistribution than most other biological specimens. Because of these properties, it has become a reliable substrate for both forensic and anti-doping research.

Analytical protocols, most commonly GC–MS and LC–MS/MS, have been adapted to the high-water, low-protein composition of VH. These allow detection of endogenous markers such as electrolytes, which support postmortem interval (PMI) estimation, and xenobiotics including ethanol, therapeutic drugs, and anabolic steroids at nanogram-per-milliliter levels. Consequently, VH is widely applied in routine forensic practice and is gaining recognition in doping control, especially in cases involving deceased athletes.

5.12.1 Physiological Basis and Collection

Vitreous humor is a transparent, gel-like fluid composed of around 98% water, with dissolved electrolytes, metabolites, and trace proteins. Its compartmentalization by the blood-retinal barrier contributes to its stability and resistance to microbial degradation after death. Collection is typically performed postmortem by sterile aspiration through the sclera, yielding 2–4 mL per eye. Once collected, samples are stored at 4 °C or frozen to preserve analyte integrity for subsequent testing.

5.12.2 Forensic Applications

VH is particularly valuable in postmortem toxicology when blood is compromised by hemolysis or putrefaction. It provides reliable evidence of substances such as ethanol, opioids, benzodiazepines, and antidepressants. Importantly, it helps differentiate between antemortem alcohol ingestion and ethanol produced postmortem by microbial activity. Biochemical markers such as potassium and hypoxanthine also serve as indicators for PMI estimation. Electrolyte imbalances or abnormal metabolite levels can reveal underlying conditions such as dehydration, diabetic ketoacidosis, or renal failure, providing pathologists with crucial diagnostic insights.

5.12.3 Doping Detection Applications

In cases of sudden athlete deaths, VH has been shown to retain doping agents such as anabolic steroids, stimulants, and β_2-agonists that may no longer be detectable in blood. This makes it invaluable for determining whether performance-enhancing drugs contributed to fatal outcomes. In veterinary settings, VH also proves useful for postmortem doping investigations, such as in horse racing, where its stability allows for reliable detection of substances including clenbuterol or corticosteroids.

5.12.4 Advantages and Limitations

Vitreous humor is resistant to decomposition and largely unaffected by postmortem redistribution. It also provides a fairly stable reflection of the body's biochemical state before death. However, it also presents limitations. Collection is invasive and restricted to postmortem settings, the sample volume is small, and concentrations of certain drugs may be low, requiring highly sensitive instrumentation.

5.12.5 Future Perspectives

Emerging innovations are set to expand VH's role in toxicology and doping investigations. Micro-sampling and ambient ionization techniques aim to reduce sample requirements, while metabolomics and proteomics promise to uncover new biomarkers beyond traditional electrolytes. Collaborative efforts by international bodies such as TIAFT are driving the standardization of collection and analysis, which is expected to improve reliability across laboratories and strengthen the evidentiary value of VH-based findings.

With its resilience against decomposition and consistent biochemical stability, vitreous humor remains an essential matrix in forensic toxicology and retrospective doping control. By combining its inherent properties with advanced analytical technologies, VH provides reliable results when blood or urine are unavailable, bridging crucial gaps in both justice and sports integrity.

5.13 Brain and Kidney Tissues

Brain and kidney tissues have emerged as vital matrices in forensic toxicology and retrospective doping studies due to their distinct physiological roles, ability to accumulate xenobiotics, and extended stability after death [47, 48]. Unlike blood or urine, which degrade rapidly postmortem, these tissues can preserve meaningful evidence for longer periods, making them particularly valuable when conventional samples are unavailable or compromised.

The brain, protected by the blood–brain barrier and characterized by its lipid-rich environment, is a key site for the retention of lipophilic drugs. This property allows forensic scientists to detect substances even after they have disappeared from circulation, thereby providing critical evidence in cases of overdose, poisoning, or concealed substance abuse [49, 50]. Conversely, the kidney plays a central role in filtration and excretion, concentrating hydrophilic compounds and their metabolites. This makes it highly informative in assessing both acute intoxication and chronic exposure to drugs, toxins, or doping agents.

5.13.1 Physiological and Toxicological Basis

The distribution of xenobiotics within these organs reflects their pharmacokinetic behavior. In the brain, the lipid-rich environment supports the accumulation of fat-soluble compounds, while the blood–brain barrier restricts rapid equilibration with blood. This creates a prolonged detection window for neuroactive substances. The kidney, in contrast, actively filters and secretes polar drugs and metabolites, often concentrating them to levels above those found in blood or urine. These unique properties extend the interpretive value of both tissues in forensic and anti-doping investigations.

5.13.2 Brain Tissue in Forensic and Doping Applications

In forensic contexts, brain tissue is indispensable for identifying neurotoxic substances such as carbon monoxide, cyanide, or heavy metals. It also provides definitive evidence in drug overdose cases, particularly with lipophilic drugs like opioids or benzodiazepines, which may persist even when blood levels are inconclusive. Chronic substance abuse, including alcoholism or methamphetamine use, can be investigated through biomarkers of neurodegeneration or structural damage detected in brain tissue.

In anti-doping investigations, the brain offers a complementary perspective by retaining psychostimulants, such as cocaine or amphetamines, which may be cleared quickly from urine. Although still experimental, research has also explored brain tissue as a potential site for detecting genetic modifications associated with performance enhancement.

5.13.3 Kidney Tissue in Forensic and Doping Applications

Kidney tissue provides equally critical insights. It can reveal evidence of nephrotoxic exposures, such as damage from ethylene glycol, heavy metals, or certain pharmaceuticals. Histopathological studies of kidney sections often identify acute tubular necrosis, fibrosis, or oxalate crystal deposition, confirming the cause of renal failure in poisoning cases. Furthermore, the kidney retains metabolites of opioids, barbiturates, and antipsychotics, making it invaluable in overdose investigations and chronic poisoning assessments.

In the doping field, kidney tissues are particularly useful for detecting diuretics, masking agents, and plasma expanders that may be overlooked in routine urine testing. Histological changes linked to anabolic steroid use, such as glomerulosclerosis, can also be observed. Additionally, renal responses to synthetic EPO may leave identifiable traces, including hyperplasia of EPO-producing cells.

5.13.4 Comparative Advantages and Challenges

Brain tissue directly links neuroactive substances to central nervous system impairment, providing a powerful means of correlating biochemical evidence with cause of death. Kidney tissue, on the other hand, serves as a long-term record of toxic exposures and structural damage, offering complementary perspectives.

Nevertheless, challenges persist, including heterogeneous drug distribution, matrix complexity, and the influence of postmortem redistribution. These issues highlight the need for standardized sampling protocols, defined anatomical reference sites, and consistent storage conditions.

5.13.5 Future Directions

The future of brain and kidney toxicology lies in integrating Mass Spectrometry Imaging (MSI) with quantitative workflows to deliver both concentration and spatial data, adopting micro-sampling probes to minimize tissue use, and applying metabolomic and proteomic strategies to identify novel biomarkers. Standardization efforts by organizations such as TIAFT will be key to enhancing reproducibility and strengthening legal defensibility.

Brain and kidney tissues provide unique and indispensable perspectives in forensic toxicology and doping investigations. They not only extend detection windows beyond traditional samples but also reveal mechanistic insights into toxicity, chronic exposure, and physiological damage. With advancing analytical technologies and harmonized protocols, their role in bridging scientific precision, legal reliability, and ethical accountability will only strengthen in the years ahead.

5.14 Bone and Bone Marrow

Bone and bone marrow have become essential alternative matrices in forensic toxicology and doping investigations, especially in cases where conventional samples such as blood, urine, or soft tissues are unavailable [51]. This often occurs in situations involving advanced decomposition or skeletonized remains, where these hard tissues remain preserved long after other biological evidence has degraded. Their unique structural and biochemical properties allow for the retention of a wide range of xenobiotics, extending detection windows well beyond those of traditional specimens.

Bone is composed primarily of hydroxyapatite and collagen, forming a dense mineralized scaffold that provides both durability and an affinity for certain compounds [52, 53]. Its relatively low perfusion rate and capacity to bind lipophilic drugs enable long-term sequestration within its mineralized structure. In contrast, bone marrow, located within the medullary cavities, is highly vascularized and rich in lipids. This environment facilitates the accumulation of both hydrophilic and lipophilic substances, reflecting systemic drug intake in a way that parallels blood plasma. Bone and bone marrow provide complementary insights, with bone often reflecting historical or cumulative exposure and bone marrow revealing more recent or active substance use.

Recent advances in analytical chemistry have transformed bone and marrow analysis. Demineralization, homogenization, and extraction techniques combined with UHPLC-MS/MS and GC–MS enable detection of drugs and metabolites at nanogram-per-gram levels. Bone marrow homogenates are particularly amenable to LC–MS for detecting opioids, steroids, and stimulants.

Specialized approaches add further depth. Laser ablation-inductively coupled plasma mass spectrometry (LA-ICP-MS) maps metal distribution in bone to assess long-term exposure to toxic elements. Fourier transform infrared spectroscopy (FTIR) can evaluate collagen degradation to help estimate postmortem intervals, while histopathology and flow cytometry of marrow reveal cellular changes linked to doping or chronic drug use.

5.14.1 Forensic Applications

In forensic science, bone and bone marrow serve multiple purposes. They are reliable sources of nuclear and mitochondrial DNA, crucial for human identification in mass disasters, archaeological cases, or criminal investigations. Stable isotope analysis of bone can also reveal geographic origins or dietary histories, aiding in reconstructing life histories of unidentified individuals.

From a toxicological perspective, bones retain heavy metals such as lead, cadmium, or arsenic, serving as markers of chronic occupational or environmental exposure. Bone marrow, protected by the mineralized matrix, preserves drugs like antidepressants, opioids, or benzodiazepines even when blood and organs are degraded. These properties make them invaluable in investigating poisoning, overdose, or long-term exposure cases.

5.14.2 Doping Detection Applications

Bone and marrow are increasingly explored in anti-doping science. Lipophilic agents such as anabolic steroids can accumulate in marrow for months, allowing retrospective detection far beyond the short detection windows of urine tests. As the site of red blood cell production, marrow may also retain biomarkers of EPO misuse. In veterinary and animal sports, marrow testing has revealed prohibited drugs in racehorses and livestock, ensuring compliance with both sporting and food safety regulations.

5.14.3 Advantages, Challenges, and Future Directions

The durability of bone makes it one of the most reliable tissues for long-term investigations, while marrow provides a valuable parallel to plasma when blood is unavailable. However, challenges remain, including labor-intensive demineralization steps, low analyte concentrations, and ethical concerns surrounding bone retention in certain cultural contexts.

Emerging technologies promise to address many of these limitations. Non-destructive imaging methods such as micro-CT preserve sample integrity while allowing structural evaluation. Metabolomic profiling of marrow may identify new biomarkers of drug exposure, while machine learning applied to bone degradation patterns could refine postmortem interval estimation. International efforts to standardize sampling sites, preparation methods, and storage protocols will be critical in enhancing comparability and legal defensibility.

5.15 Cerebrospinal Fluid (CSF)

CSF is an important but underutilized biological matrix in forensic toxicology and retrospective anti-doping research. Enclosed within the central nervous system and protected by the blood–brain barrier, CSF offers a chemically stable and relatively isolated medium compared to blood or urine. Its low protein concentration, minimal enzymatic activity, and resistance to postmortem redistribution make it particularly valuable in situations where conventional samples are degraded [54]. In forensic practice, CSF has proven especially useful for determining ethanol concentrations when blood is compromised by putrefaction, hemolysis, or microbial activity. Furthermore, it provides direct insights into compounds that actively cross into the central nervous system, thereby shedding light on the timing of drug exposure and its potential effects on neurological function [55, 56].

5.15.1 Physiological and Toxicokinetic Basis

The composition of CSF reflects both passive diffusion across the blood CSF barrier and active transport mechanisms occurring at the choroid plexus. Because the matrix is acellular and contains little protein, lipophilic drugs and their metabolites are more likely to remain in free solution rather than binding to proteins or undergoing enzymatic degradation. This enhances their detectability in postmortem examinations. In ethanol-related fatalities, CSF ethanol levels correlate strongly with antemortem blood concentrations, making CSF a reliable alternative when blood analysis is unreliable. Similarly, CSF sampling can highlight the presence of central nervous system–active drugs, such as opioids or benzodiazepines, with minimal interference from redistribution artifacts.

Due to its relatively clean matrix, CSF requires less extensive preparation than blood or tissue. Common approaches include protein precipitation and solid-phase extraction, followed by chromatographic separation coupled with mass spectrometric detection. Techniques such as LC–MS/MS and GC–MS offer highly sensitive, quantitative results at low milligram-per-liter levels. Importantly, sample integrity must be preserved by avoiding contamination with blood during collection, as even minor hemolysis can distort analyte profiles.

5.15.2 Forensic Toxicology Applications

CSF analysis has been successfully applied to medico-legal investigations involving intoxication and cause of death. It is particularly useful when peripheral blood and urine are unavailable or severely degraded. Drugs such as opioids, benzodiazepines, and ethanol can be reliably quantified in CSF, and the stability of this matrix also allows exploration of biochemical markers urea, creatinine, and electrolytes that may support postmortem interval estimation. In addition, CSF can reveal novel psychoactive substances or highly potent opioids that may not be detectable in peripheral fluids due to extensive metabolism or instability.

5.15.3 Doping Detection Applications

Although not commonly used in live anti-doping programs, CSF shows promise for postmortem investigations involving performance-enhancing substances. Because many stimulants and neuroactive peptides act directly on the central nervous system, CSF analysis provides unique evidence of their use. Studies have demonstrated that amphetamines and similar compounds may be detectable in CSF for longer periods than in plasma, making it a valuable matrix in delayed investigations. In veterinary settings, CSF testing has been suggested for detecting illegal administration of CNS-active agents in livestock and competitive animals, where drug distribution may differ significantly from blood.

5.15.4 Challenges and Future Directions

Despite its advantages, CSF analysis presents several challenges. Collection requires specialized autopsy techniques, and contamination from blood remains a risk if the meninges are punctured improperly. Harmonization of sampling methods, particularly between lumbar and ventricular sites, is needed to ensure consistency across laboratories. Looking ahead, metabolomic and proteomic profiling of CSF may expand its role by identifying novel biomarkers of toxic exposure and peptide-based doping agents. Advances in microfluidic technologies and imaging-guided sampling are likely to improve recovery while minimizing contamination. Furthermore, machine-learning models trained on multi-omic CSF datasets could enhance both toxicological interpretation and postmortem interval estimation, solidifying CSF as a critical matrix in forensic and anti-doping sciences.

5.16 Summary

This chapter explored the critical role of diverse biological matrices in forensic toxicology and anti-doping investigations, emphasizing their unique physiological properties, analytical considerations, and practical applications. Traditional matrices such as blood and urine remain the backbone of toxicological analysis, providing rapid and reliable insights into recent substance use. However, alternative matrices including breath, vitreous humor, cerebrospinal fluid, liver, kidney, brain, bone, bone marrow, gastric contents, and specialized veterinary samples offer complementary or extended detection windows, particularly in postmortem or complex investigative scenarios.

Analytical advancements, including high-resolution chromatography, mass spectrometry imaging, microextraction, and metabolomic profiling, have significantly improved sensitivity, specificity, and spatial resolution across all matrices. The chapter also addressed challenges such as matrix heterogeneity, postmortem redistribution, and the need for standardized collection and storage protocols to ensure reliability and legal defensibility.

Overall, this chapter highlighted how integrating traditional and alternative biological matrices enhances forensic investigations, enabling accurate reconstruction of substance exposure, determination of cause of death, and retrospective doping detection. The evolving analytical approaches continue to expand the scope and reliability of toxicological assessments, bridging the gap between scientific rigor, legal accountability, and ethical considerations in forensic and anti-doping practices.

References

1. Jones S, McGowan C, Boyle S, et al. An overview of sample preparation in forensic toxicology. Wiley Interdiscip Rev Forensic Sci. 2022;4: e1436. https://doi.org/10.1002/wfs2.1436.
2. de Campos EG, da Costa BRB, Dos Santos FS, et al. Alternative matrices in forensic toxicology: a critical review. Forensic Toxicol. 2022;40:1–18. https://doi.org/10.1007/s11419-021-00596-5.
3. Jurowski K, Noga M, Krośniak A, et al. Recommendations, trends and analytical strategies applied for biological samples collection (sampling) and storage in forensic toxicology of

volatile poisons (inorganic and organic). TrAC Trends Anal Chem. 2024;181(Part A):118033. https://doi.org/10.1016/j.trac.2024.118033
4. Dinis-Oliveira RJ, Carvalho F, Duarte JA, et al. Collection of biological samples in forensic toxicology. Toxicol Mech Methods. 2010;20:363–414. https://doi.org/10.3109/15376516.2010.497976.
5. Penetar DM, McNeil JF, Ryan ET, et al. Comparison among plasma, serum, and whole blood ethanol concentrations: impact of storage conditions and collection tubes. J Anal Toxicol. 2008;32:505–10. https://doi.org/10.1093/jat/32.7.505.
6. Arce-López B, Lizarraga E, Vettorazzi A, et al. Human biomonitoring of mycotoxins in blood, plasma and serum in recent years: a review. Toxins. 2020;12:147. https://doi.org/10.3390/toxins12030147.
7. Schulz M, Iwersen-Bergmann S, Andresen H, et al. Therapeutic and toxic blood concentrations of nearly 1,000 drugs and other xenobiotics. Crit Care. 2012;16:R136. https://doi.org/10.1186/cc11441.
8. Faiss R, Saugy J, Saugy M. Fighting doping in elite sports: blood for all tests! Front Sports Act Living. 2019;1:30. https://doi.org/10.3389/fspor.2019.00030
9. Bickel J, Jungen H, Müller A, et al. Toxicologic analysis of 35 drugs in post mortem human blood samples with focus on antihypertensive and antiarrhythmic drugs. J Chromatogr B Analyt Technol Biomed Life Sci. 2024;1242: 124196. https://doi.org/10.1016/j.jchromb.2024.124196.
10. Anderson NL, Anderson NG. The human plasma proteome: history, character, and diagnostic prospects. Mol Cell Proteomics. 2002;1:845–67. https://doi.org/10.1074/mcp.r200007-mcp200.
11. Gjerde H, Oyono Y, Jamt REG, et al. Drug analysis: comparison between dried plasma spots and liquid plasma samples of trauma patients from Cameroon-a feasibility study. Drug Test Anal. 2024;16:303–8. https://doi.org/10.1002/dta.3545.
12. Meyer GM, Maurer HH, Meyer MR. Multiple stage MS in analysis of plasma, serum, urine and in vitro samples relevant to clinical and forensic toxicology. Bioanalysis. 2016;8:457–81. https://doi.org/10.4155/bio.16.15.
13. Psychogios N, Hau DD, Peng J, et al. The human serum metabolome. PLoS ONE. 2011;6: e16957. https://doi.org/10.1371/journal.pone.0016957.
14. Wissenbach DK, Steuer AE. Advances in testing for sample manipulation in clinical and forensic toxicology—Part A: urine samples. Anal Bioanal Chem. 2023;415:5101–15. https://doi.org/10.1007/s00216-023-04711-w.
15. Ceelen M, Dorn T, Buster M, et al. Post-mortem toxicological urine screening in cause of death determination. Hum Exp Toxicol. 2010;30:1165–73. https://doi.org/10.1177/0960327110390063.
16. Tenore PL. Advanced urine toxicology testing. J Addict Dis. 2010;29:436–48. https://doi.org/10.1080/10550887.2010.509277.
17. Schneider TD, Binz TM, Kraemer T, et al. New approaches to identify urine and hair adulteration attempts in forensic toxicology: a proof-of-concept study using a proteomics approach based on liquid chromatography-mass spectrometry (LC-MS). Anal Chem. 2025;97:17761–9. https://doi.org/10.1021/acs.analchem.5c03096.
18. Kintz P. Hair analysis in forensic toxicology: an updated review with a special focus on pitfalls. Curr Pharm Des. 2017;23:5480–6. https://doi.org/10.2174/1381612823666170929155628.
19. Jurado C. Forensic applications of hair analysis. In: Kintz P, Salomone A, Vincenti M (eds) Hair Analysis in Clinical and Forensic Toxicology, pp. 241–273. Academic Press, London (2015). https://doi.org/10.1016/B978-0-12-801700-5.00009-1
20. Ferreira C, Paulino C, Quintas A. Extraction procedures for hair forensic toxicological analysis: a mini-review. Chem Res Toxicol. 2019;32:2367–81. https://doi.org/10.1021/acs.chemrestox.9b00301.
21. Usman M, Naseer A, Baig Y, et al. Forensic toxicological analysis of hair: a review. Egypt J Forensic Sci. 2019;9:17. https://doi.org/10.1186/s41935-019-0119-5.
22. Drummer OH. Drug testing in oral fluid. Clin Biochem Rev. 2006;27:147–59.

23. Harper CE, Hudson JS, Tidwell K, et al. Implementation of the first comprehensive state oral fluid drug testing program for roadside screening and laboratory testing in DUID cases-a 5-year review. J Anal Toxicol. 2023;47:694–702. https://doi.org/10.1093/jat/bkad051.
24. De La Torre R, Pichini S. Sweat: toxicology. In: Jamieson A, Moenssens A (eds) Wiley Encyclopedia of Forensic Science. Wiley, Hoboken (2009). https://doi.org/10.1002/9780470061589.fsa412
25. de la Torre Fornell R, Pichini S. Alternative specimens-sweat. In: Jamieson A, Moenssens A (eds) Wiley Encyclopedia of Forensic Science. Wiley, Hoboken (2014). https://doi.org/10.1002/9780470061589.fsa412.pub2
26. Koster RA, Alffenaar JW, Greijdanus B, et al. Application of sweat patch screening for 16 drugs and metabolites using a fast and highly selective LC-MS/MS method. Ther Drug Monit. 2014;36:35–45. https://doi.org/10.1097/FTD.0b013e3182a04feb.
27. Chawarski MC, Fiellin DA, O'Connor PG, et al. Utility of sweat patch testing for drug use monitoring in outpatient treatment for opiate dependence. J Subst Abuse Treat. 2007;33:411–5. https://doi.org/10.1016/j.jsat.2007.03.004.
28. Moore C, Negrusz A, Lewis D. Determination of drugs of abuse in meconium. J Chromatogr B Analyt Technol Biomed Life Sci. 1998;713:137–46. https://doi.org/10.1016/s0378-4347(97)00479-9.
29. Price HR, Collier AC, Wright TE. Screening pregnant women and their neonates for illicit drug use: consideration of the integrated technical, medical, ethical, legal, and social issues. Front Pharmacol. 2018;9:961. https://doi.org/10.3389/fphar.2018.00961.
30. Academy of Breastfeeding Medicine Protocol Committee, Jansson LM. ABM clinical protocol #21: guidelines for breastfeeding and the drug-dependent woman. Breastfeed Med. 2009;4:225–228.https://doi.org/10.1089/bfm.2009.9987
31. Busardò FP, Bertol E, Mannocchi G, et al. Determination of GHB levels in breast milk and correlation with blood concentrations. Forensic Sci Int. 2016;265:172 81. https://doi.org/10.1016/j.forsciint.2016.02.020.
32. Yang Q, Chen D, Liu X, et al. Identification of nanoparticle infiltration in human breast milk: chemical profiles and trajectory pathways. Proc Natl Acad Sci USA. 2025;122: e2500552122. https://doi.org/10.1073/pnas.2500552122.
33. Solimini R, Minutillo A, Kyriakou C, et al. Nails in forensic toxicology: an update. Curr Pharm Des. 2017;23:5468–79. https://doi.org/10.2174/1381612823666170704123126.
34. Baumgartner MR. Nails: an adequate alternative matrix in forensic toxicology for drug analysis? Bioanalysis. 2014;6:2189–91. https://doi.org/10.4155/bio.14.165.
35. Moura PC, Raposo M, Vassilenko V. Breath volatile organic compounds (VOCs) as biomarkers for the diagnosis of pathological conditions: a review. Biomed J. 2023;46: 100623. https://doi.org/10.1016/j.bj.2023.100623.
36. Hanna GB, Boshier PR, Markar SR, et al. Accuracy and methodologic challenges of volatile organic compound-based exhaled breath tests for cancer diagnosis: a systematic review and meta-analysis. JAMA Oncol. 2019;5: e182815. https://doi.org/10.1001/jamaoncol.2018.2815.
37. Thevis M, Krug O, Geyer H, et al. Expanding analytical options in sports drug testing: mass spectrometric detection of prohibited substances in exhaled breath. Rapid Commun Mass Spectrom. 2017;31:1290–6. https://doi.org/10.1002/rcm.7903.
38. Thevis M, Geyer H, Tretzel L, et al. Sports drug testing using complementary matrices: advantages and limitations. J Pharm Biomed Anal. 2016;130:220–30. https://doi.org/10.1016/j.jpba.2016.03.055.
39. Ketola RA, Kriikku P. Drug concentrations in post-mortem specimens. Drug Test Anal. 2019;11:1338–57. https://doi.org/10.1002/dta.2662.
40. Schneider J, Mas-Carrió E, Jan C, et al. Comprehensive coverage of human last meal components revealed by a forensic DNA metabarcoding approach. Sci Rep. 2021;11:8876. https://doi.org/10.1038/s41598-021-88418-x.
41. Orfanidis A, Gika H, Theodoridis G, et al. Development of a UHPLC-MS/MS method for the determination of 84 pharmaceuticals and drugs of abuse in human liver. J Chromatogr B Analyt Technol Biomed Life Sci. 2020;1151: 122192. https://doi.org/10.1016/j.jchromb.2020.122192.

42. Legge CM, Payne-James JJ, Puntis JWL, et al. Postmortem gastric content analysis. In: Gall JAM, Payne-James JJ (eds) Current Practice in Forensic Medicine. Wiley, Hoboken (2016). https://doi.org/10.1002/9781118456026.ch3
43. Peres MD, Nascimento S, Pelição FS. A new clean-up approach for stomach content toxicological analysis. Forensic Sci Int. 2019;302: 109936. https://doi.org/10.1016/j.forsciint.2019.109936.
44. Bierly J, Labay LM. The utility of bile in postmortem forensic toxicology. Acad Forensic Pathol. 2018;8:324–7. https://doi.org/10.1177/1925362118782073.
45. Bévalot F, Cartiser N, Bottinelli C, et al. Vitreous humor analysis for the detection of xenobiotics in forensic toxicology: a review. Forensic Toxicol. 2016;34:12–40. https://doi.org/10.1007/s11419-015-0294-5.
46. Li W, Chang Y, Cheng Z, et al. Vitreous humor: a review of biochemical constituents in postmortem interval estimation. J Forensic Sci Med. 2018;4:85–90. https://doi.org/10.4103/jfsm.jfsm_13_18.
47. Hansen SL, Dalsgaard PW, Linnet K, et al. Comparison of comprehensive screening results in postmortem blood and brain tissue by UHPLC-QTOF-MS. J Anal Toxicol. 2023;46:1053–8. https://doi.org/10.1093/jat/bkac011.
48. Rohrig TP, Hicks CA. Brain tissue: a viable postmortem toxicological specimen. J Anal Toxicol. 2015;39:137–9. https://doi.org/10.1093/jat/bku139.
49. Uekusa K, Hayashida M, Ohno Y. Forensic toxicological analyses of drugs in tissues in formalin solutions and in fixatives. Forensic Sci Int. 2015;249:165–72. https://doi.org/10.1016/j.forsciint.2015.01.028.
50. Steuer AE, Brockbals L, Poetzsch SN, et al. Evaluation of bladder wash as an alternative specimen for postmortem toxicology: comparison to screening results of urine and kidney tissue. Drug Test Anal. 2025;17:1640–55. https://doi.org/10.1002/dta.3861.
51. Rubin KM. The current state and future directions of skeletal toxicology: forensic and humanitarian implications of a proposed model for the in vivo incorporation of drugs into the human skeleton. Forensic Sci Int. 2018;289:419–28. https://doi.org/10.1016/j.forsciint.2018.06.024.
52. Giordano G, Biehler-Gomez L, Cattaneo C, et al. Toxicological analyses of the bone matrix: successes and challenges. Archaeometry. 2025;67:1001–21. https://doi.org/10.1111/arcm.13065.
53. Vandenbosch M, Rooseleers L, Van Den Bogaert et al. Skeletal tissue, a viable option in forensic toxicology? A view into post mortem cases. Forensic Sci Int. 2020;309:110225. https://doi.org/10.1016/j.forsciint.2020.110225
54. Bohnert S, Reinert C, Trella S, et al. Neuroforensomics: metabolites as valuable biomarkers in cerebrospinal fluid of lethal traumatic brain injuries. Sci Rep. 2024;14:13651. https://doi.org/10.1038/s41598-024-64312-0.
55. Wachholz P, Skowronek R, Pawlas N. Assessing the applicability of cerebrospinal fluid collected from the spinal cord for the determination of ethyl alcohol in post-mortem toxicology. Forensic Sci Med Pathol. 2023;19:44–9. https://doi.org/10.1007/s12024-022-00560-8.
56. Tominaga M, Michiue T, Ishikawa T, et al. Evaluation of postmortem drug concentrations in cerebrospinal fluid compared with blood and pericardial fluid. Forensic Sci Int. 2015;254:118–25. https://doi.org/10.1016/j.forsciint.2015.07.005.

Chapter 6
Extraction Techniques in Forensic Toxicology

In this chapter, a comprehensive overview of extraction techniques employed in forensic toxicology and anti-doping analysis is presented. Accurate detection of drugs, metabolites, and prohibited substances from complex biological matrices such as blood, urine, hair, and oral fluid depends largely on the effectiveness of sample preparation. The chapter systematically discusses conventional and modern extraction techniques, highlighting their principles, advantages, limitations, and applications in both forensic casework and sports drug testing. LLE remains a widely applied classical approach due to its simplicity and cost-effectiveness, yet it often requires large solvent volumes and may lack selectivity. Solid-phase extraction (SPE) offers improved cleanup and reproducibility, while solid-phase microextraction (SPME) introduces solvent-free sampling with minimal sample handling. More advanced miniaturized techniques, such as dispersive liquid–liquid microextraction (DLLME), combine high enrichment factors with reduced solvent consumption. Similarly, stir bar sorptive extraction (SBSE) enhances analyte recovery for trace-level detection, particularly in environmental and biological applications. Emerging methods like supported liquid extraction (SLE) and magnetic solid-phase extraction (MSPE) demonstrate significant progress in automation, reusability, and environmental sustainability, offering higher sensitivity for complex toxicological and doping analyses. The dilute-and-inject technique, although simple, is limited to clean matrices and often requires high-performance instruments to tolerate direct sample loading. By comparing the strengths and drawbacks of each method, this chapter emphasizes that no single technique is universally superior. Instead, method selection depends on analyte properties, matrix complexity, available resources, and the analytical goal. Together, these extraction strategies form the foundation of reliable toxicological interpretation and fair enforcement in anti-doping practices, ensuring scientific accuracy and ethical integrity.

Forensic toxicology · Anti-doping analysis · Sample preparation techniques · LLE · Solid-Phase Extraction (SPE) · Solid-Phase Microextraction (SPME) · Dispersive Liquid–Liquid Microextraction (DLLME) · Stir Bar Sorptive Extraction

A. K. Karakka Kal et al., *Advanced Extraction Techniques for Drug and Toxin Detection*, https://doi.org/10.1007/978-3-032-17335-5_6

(SBSE) · Magnetic Solid-Phase Extraction (MSPE) · Supported Liquid Extraction (SLE)

6.1 Overview

Extraction techniques are fundamental to forensic toxicology and anti-doping science, serving as the first and often most critical step in the analytical workflow. Their role is to isolate, purify, and concentrate target compounds from complex biological matrices so that they can be accurately detected and quantified. Inefficiencies at this stage can compromise the entire analysis. For example, in LC–MS/MS, variations introduced during sample preparation are among the leading causes of inconsistent results [1]. Every additional preparation step adds a potential source of error, which can increase analytical uncertainty.

One of the greatest challenges in sample preparation is overcoming matrix effects, particularly those caused by endogenous phospholipids. These compounds frequently contribute to ion suppression in LC–MS/MS, reducing both reproducibility and sensitivity. In addition, the pharmacokinetic properties of many lipophilic drugs complicate their analysis. Certain synthetic cannabinoids, for instance, display extremely high plasma protein binding values (90.8%–99.9%), which prolong their clearance and complicate detection in toxicological testing. Such challenges underscore how closely extraction efficiency is tied to the accuracy of forensic and anti-doping analyses [2]. Table 6.1 summarizes the plasma protein binding characteristics of major drug categories, highlighting their relevance in both therapeutic monitoring and forensic analysis.

Applications in doping control, postmortem toxicology, and workplace testing highlight the critical importance of extraction methods [3]. In doping control, the WADA requires laboratories to adhere to strict sensitivity thresholds, which can only be achieved with robust extraction procedures [4]. Postmortem investigations often involve decomposed specimens, where efficient methods are vital for reliable identification of drugs and toxins. In workplace drug testing, the emphasis lies on chain-of-custody integrity, yet extraction remains equally crucial to avoid false positives or negatives. Recent advances demonstrate that integrating SPE with LLE into hybrid workflows can markedly improve recovery and detection, particularly in hair analysis where sensitivity is often a limiting factor.

The diversity of biological specimens also necessitates tailored preparation strategies. In urine, many drugs are excreted as glucuronidated metabolites, requiring enzymatic hydrolysis with β-glucuronidase at carefully optimized pH and temperature conditions. Blood samples, by contrast, demand efficient protein removal, often achieved through organic solvents such as acetonitrile, to reduce interference in downstream analysis. Hair samples require sequential washing with solvents like isopropanol or phosphate buffer to remove external contamination, ensuring that results reflect systemic rather than environmental exposure. For oral fluid, stabilization buffers are commonly added to protect analytes from enzymatic degradation, preserving sample integrity until analysis [5].

Equally important is the chemical nature of analytes, which dictates the most effective extraction conditions. Basic drugs such as amphetamines are best recovered under alkaline conditions, often using mixed-mode cation-exchange solid-phase extraction. Acidic compounds, including many NSAIDs, require acidic conditions and specially designed cartridges to ensure retention and elution [6]. Steroids typically require enzymatic deconjugation before extraction, which is often carried out with C18-based techniques. Polar metabolites pose another challenge and benefit from hydrophilic interaction liquid chromatography (HILIC) optimized protocols that enhance retention and separation. Matching preparation methods to both the specimen type and drug class is essential for accuracy [7, 8].

In forensic toxicology, validated extraction methods are indispensable because they directly influence key performance metrics [9]. According to the ICH M10 guidelines adopted by both the FDA and EMA, method validation requires acceptable recovery rates, minimal matrix effect suppression, and appropriately defined limits of quantification [10]. Forensic laboratories typically target recovery values within reproducible ranges, matrix effects under 15% variation, and detection limits that align with the analyte's biological relevance. These parameters ensure both the reliability and defensibility of results in judicial and regulatory contexts.

Another challenge is the low concentrations at which many analytes are present often measured in parts per million (ppm), billion (ppb), or even trillion (ppt). Such

Table 6.1 Comparative Plasma Protein Binding (PPB) of common therapeutic and illicit drugs

Drug class	Example drugs	PPB (%)
Antipsychotics	Haloperidol, Clozapine	90–98
Anticoagulants	Warfarin	~99
Antiretrovirals	Efavirenz, Ritonavir	99
Antibiotics	Ceftriaxone	85–95
Antiepileptics	Phenytoin, Valproic Acid	90–95
Antidepressants	Amitriptyline, Sertraline	90–95
Beta-blockers	Propranolol, Metoprolol	90–95
Benzodiazepines	Diazepam, Lorazepam	85–99
Calcium channel blockers	Verapamil, Diltiazem	90–95
Opioids	Fentanyl, Methadone	80–90
NSAIDs	Ibuprofen, Naproxen	95–99
Statins	Atorvastatin, Simvastatin	95–99

Notes The PPB values indicate the percentage of the drug bound to plasma proteins, primarily albumin. High PPB can influence the drug's distribution, elimination, and potential for drug–drug interactions. These values are approximate and can vary based on individual patient factors and specific formulations levels may fall below the detection thresholds of even the most advanced analytical instruments. To address this, extraction techniques must also function as preconcentration tools, enriching trace analytes while removing contaminants that could interfere with analysis. Advances in microextraction and hybrid workflows are particularly promising in this regard, as they offer high sensitivity while reducing solvent consumption and sample handling steps.

Overall, the evolution of extraction methods in forensic toxicology reflects a balance between practicality, sensitivity, and reproducibility. From traditional approaches such as LLE to modern miniaturized techniques like dispersive liquid–liquid microextraction, supported liquid extraction, and magnetic solid-phase extraction, the goal remains the same: to provide clean, concentrated samples that yield accurate and reliable results [11, 12]. Continued innovation in this field not only improves detection capabilities but also enhances the scientific integrity of forensic and anti-doping investigations, ensuring that outcomes are both trustworthy and defensible.

Key extraction methods include

Liquid–Liquid Extraction (LLE): This traditional method separates compounds based on their differential solubility in two immiscible liquids, typically an aqueous phase and an organic solvent. While LLE is straightforward and widely utilized, it can be time-consuming and often requires large volumes of solvents [13].

Solid-Phase Extraction (SPE): SPE employs a solid sorbent material to isolate analytes from liquid samples. It offers advantages over LLE, such as reduced solvent usage and enhanced selectivity. However, SPE can be labor-intensive and may involve higher costs due to the need for specialized equipment [14].

Solid-Phase Microextraction (SPME): SPME integrates sampling, extraction, and concentration into a single step using a coated fiber to adsorb analytes. This solvent-free method is efficient and sensitive, making it suitable for detecting trace levels of substances [15].

Dispersive Liquid–Liquid Microextraction (DLLME): DLLME is a miniaturized LLE technique where a mixture of extraction and disperser solvents is rapidly injected into an aqueous sample, forming a cloudy solution that facilitates analyte extraction. This method is known for its simplicity, speed, and high enrichment factors [16].

Microextraction By Packed Sorbent (MEPS): MEPS is a miniaturized SPE technique that uses a small amount of sorbent packed into a syringe or pipette tip. It allows for the extraction of analytes from small sample volumes and can be automated for high-throughput analysis [17, 18].

Stir Bar Sorptive Extraction (SBSE): SBSE utilizes a stir bar coated with a sorbent to extract analytes while stirring the sample. This method is highly sensitive and suitable for extracting non-polar and moderately polar compounds from large sample volumes [19, 20].

Supported Liquid Extraction (SLE): SLE, involves passing the aqueous sample through a column containing an inert support coated with a water-immiscible solvent. It simplifies the extraction process and reduces emulsion formation compared to traditional LLE [21].

Magnetic Solid-Phase Extraction (MSPE): MSPE uses magnetic nanoparticles as sorbents to extract analytes. After extraction, an external magnetic field is applied to separate the sorbent from the sample solution, facilitating easy collection and minimizing sample handling [22].

Dilute-And-Inject: Dilute-and-inject is a simple sample preparation technique that involves the dilution of a sample followed by direct injection into an analytical instrument. A small aliquot of the sample is diluted with a suitable diluent, such as a solvent or buffer, with the purpose of reducing the effect of the interfering matrix and analyte concentration [23, 24].

Each of these techniques offers distinct advantages and may be selected based on the specific requirements of the forensic analysis, including the nature of the analyte, sample matrix, and desired sensitivity.

6.2 Liquid–Liquid Extraction (LLE)

LLE, also known as solvent extraction or partitioning, is one of the most widely applied separation techniques in analytical sciences, including forensic toxicology. Before exploring chromatographic methods, it is useful to consider LLE, as both share common underlying principles of separation [25]. In its simplest form, LLE utilizes two immiscible liquid phases most commonly a polar aqueous phase (such as water) and a non-polar organic solvent. When a sample mixture is introduced, the individual compounds distribute themselves between the two phases according to their relative solubility. This differential partitioning enables the selective isolation of target analytes from interfering substances, laying the foundation for more advanced separation and purification strategies.

6.2.1 Key Parameter of Relative Solubility

In LLE, solute distribution between two immiscible phases is described by two key parameters: the distribution coefficient (DC) and the partition coefficient (DM). The distribution coefficient expresses the concentration ratio of a solute in the organic phase relative to the aqueous phase at equilibrium. While this provides useful insight into relative solubilities, it does not always represent the absolute amount extracted when the phase volumes differ.

$$DC = \frac{solute_{org}}{solute_{aq}} = \frac{mol_{org}/V_{org}}{mol_{aq}/V_{aq}}.$$

The partition coefficient, by contrast, reflects the molar ratio of solute present in the two phases. This value provides a more direct measure of extraction efficiency, as higher DM values indicate greater transfer of the analyte into the organic phase.

$$DM = \frac{mol_{org}}{mol_{aq}}.$$

The two parameters are mathematically related, with the distribution coefficient representing the partition coefficient adjusted for the relative volumes of the two phases.

The relationship between DC and DM is:

$$DC = DM\left(\frac{V_{aq}}{V_{org}}\right).$$

In practice, multiple sequential extractions with fresh solvent are commonly performed to maximize recovery, a process that conceptually parallels the mechanism of chromatographic separations. The efficiency of solute distribution is influenced by factors such as the solute's chemical nature, solvent polarity, temperature, and pH of the system [26]. The general procedure involves the following steps:

- Solvent Selection: Choose two immiscible solvents where the target analyte has differing solubilities. Typically, this involves pairing a polar solvent like water with a non-polar organic solvent.
- Mixing*:* Combine the sample solution with the selected extraction solvent, facilitating the transfer of the analyte into the extraction solvent.
- Phase Separation: Allow the two liquid phases to separate, typically using a separatory funnel, enabling the isolation of the solvent containing the extracted analyte.
- Collection*:* Carefully collect the phase enriched with the analyte for subsequent analysis.

As illustrated in Fig. 6.1, LLE separates analytes based on their differential solubility in immiscible phases.

6.2.2 Enhancing LLE Performance

The efficiency of LLE can be significantly improved by carefully controlling several key parameters [27].

Solvent Selection

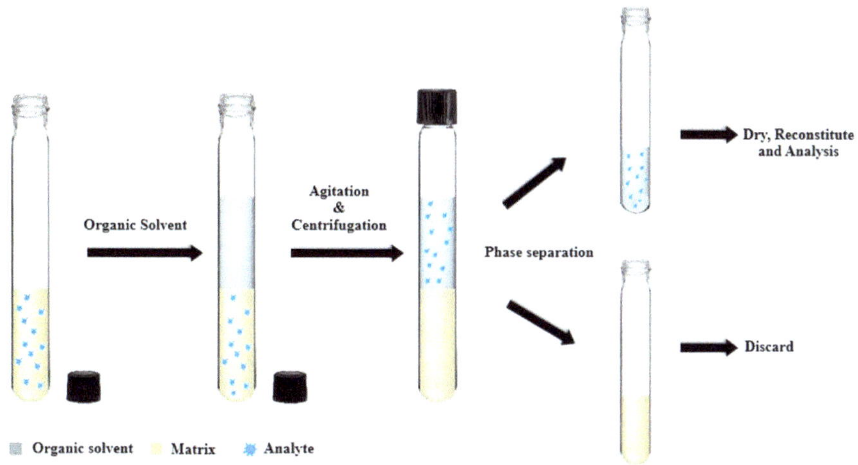

Fig. 6.1 Workflow illustration of liquid–liquid extraction

- Choose solvents that enhance analyte distribution into the organic phase.
- Aim to minimize the co-extraction of interfering impurities.

pH Control

- Modify the aqueous phase pH to control the ionization state of analytes.
- Improves partitioning efficiency into the organic solvent.

Temperature Control

- Conduct extractions at conditions that maintain analyte stability.
- Optimal temperatures can improve solubility and extraction efficiency.

Multiple Extractions

- Perform sequential extractions with fresh solvent.
- Increases overall recovery, especially for low-concentration analytes.

Solvent Selection: Selecting the appropriate extraction solvent is one of the most decisive steps in optimizing separation efficiency. The goals are to maximize the distribution coefficient (DC) of the target analyte, minimize the co-extraction of interfering compounds, and maintain solvent immiscibility along with practical handling properties.

Polarity Considerations: Polarity remains the primary factor governing extraction performance. The relationship between analyte polarity and solvent properties can be evaluated systematically. Snyder's polarity index (P') provides a practical guide for solvent selection. Non-polar analytes ($\log P > 5$) are efficiently extracted with solvents such as hexane (P' 0.1), moderately polar compounds with solvents like ethyl acetate ($P' = 4.4$), and polar metabolites ($\log P < 1$) with methanol ($P' = 5.1$). Functional

Table 6.2 Polarity index of common organic solvents

Solvent	Polarity index (P′)	Applications
Diethyl ether	2.8	Neutral drugs, amphetines
Chloroform	4.1	Basic drugs (pH $\geq$ 9)
Ethyl acetate	4.4	Broad spectrum (pH-dependent)
n-Hexane	0.1	Lipophilic compounds (steroids, THC)

group compatibility must also be considered; hydroxyl-bearing compounds favor hydrogen-bond acceptors such as ethyl acetate, aromatic systems are effectively extracted with π-interacting solvents like toluene, and ionizable groups require pH-adjusted solvent systems. As summarized in Table 6.2, Snyder's polarity index offers practical guidance for selecting solvents based on analyte polarity.

A more nuanced approach involves three-dimensional polarity assessment, taking into account dipole moment (μ), hydrogen-bonding capacity (α, β), and polarizability (π^*). Advanced polarity metrics such as the ET (30) scale, Kamlet–Taft parameters, and Hansen solubility parameters further refine solvent selection, especially when addressing novel analytes or complex biological matrices. Table 6.3 outlines a systematic approach to matching solvent polarity with analyte properties to improve extraction outcomes.

Dielectric Constant (ε): The dielectric constant (ε), or relative permittivity, is another essential property influencing solvent behavior. Solvent classification by ε values provides insight into their capacity to stabilize charged species. According to Table 6.4, solvents can be categorized by dielectric constant, which directly impacts their ability to stabilize charged species.

Mixed solvent systems are often employed to create tailored dielectric environments for example, a 7:3 mixture of hexane and ethyl acetate provides an intermediate ε of about 3.5, improving extraction selectivity. Temperature also influences ε, as rising temperatures generally reduce molecular alignment, leading to decreased permittivity. Additionally, additives can fine-tune solvent characteristics; the incorporation of just 5% methanol can raise a solvent system's dielectric constant by 15–20% due to enhanced dipole interactions and hydrogen bonding. As shown in Tables 6.5

Table 6.3 Practical framework for polarity-based solvent selection

Analytes	log P	Primary	Alternative	Additive
Non-polar	>4	n-Hexane (P′= 0.1)	Cyclohexane (P′ = 0.2)	1% triethylamine (for basic compounds)
Moderately polar	2–4	Dichloromethane (P′= 3.1)	Ethyl acetate (P′ = 4.4)	5% isopropanol (to prevent emulsion)
Polar	<2	Chloroform (P′ = 4.1)	Dichloromethane: isopropanol (9:1)	Ion-pair reagents (for charged species)

and 6.6, modern polarity measures such as ET (30), Kamlet–Taft parameters, and Hansen solubility parameters provide refined solvent selection tools.

Statistical relationships reinforce the importance of ε in solvent selection. It demonstrates a strong correlation with Snyder's polarity index ($R^2 = 0.89$), shows an inverse relationship with log P values across homologous series, and directly affects solvation energy as predicted by Born's equation, where higher dielectric media provide greater stabilization of charged solutes. Through this integrated framework considering polarity, dielectric constant, and advanced solvent parameters analysts can design extraction systems that maximize efficiency while minimizing matrix effects, thereby enhancing the reliability of forensic and anti-doping analyses.

Table 6.4 Classification of solvents by dielectric constant (ε) values

Solvent	Polarity index (P′)	Applications
Non-polar	<5	Optimal for hydrophobic compounds
• Hexane	1.9	Ideal for aliphatic hydrocarbons
• Toluene	2.4	Effective for aromatic compounds
Moderate polarity	5–20	Versatile extraction range
• Dichloromethane	8.8	Broad-spectrum applications
• Ethyl acetate	6.0	Balanced polarity
High polarity	>20	Specialized applications
• Methanol	32.7	Polar metabolites
• Water	80.1	Rarely used as organic phase

Table 6.5 Dielectric constant dependent extraction optimization

Compounds	log P	Dielectric constant (ε)	Systems
Non-polar	>4	<5	Hexane/Cyclohexane
Polar	<2	>15	Chloroform/Methanol
Zwitterionic		1 → 15	

Table 6.6 Dielectric constant-based practical implementation

Analyte class	Optimal ε Range	Recommended solvent systems
Aliphatic hydrocarbons	1.8–2.5	n-Hexane, petroleum ether
Aromatic compounds	2.4–3.0	Toluene, xylene
Halogenated compounds	8–10	Dichloromethane, chloroform
Oxygenated species	5–8	Ethyl acetate, MTBE
Nitrogen-containing	10–15	Chloroform:isopropanol (4:1)
Polar metabolites	15–25	Dichloromethane:methanol (9:1)

pH Control: The control of pH in the aqueous phase is a central factor in LLE, as it directly determines the ionization state of acidic, basic, and amphoteric analytes. Since ionization strongly influences solubility, adjusting pH effectively governs the distribution of compounds between aqueous and organic layers. A carefully controlled pH not only enhances selectivity but also improves analyte recovery and contributes to higher extract purity outcomes that are particularly important in forensic and anti-doping studies where trace-level detection is often required.

Ionization and Solubility Behavior

- For acidic analytes, ionization increases as the pH rises above the pKa value, leading to greater solubility in the aqueous phase. Conversely, lowering the pH keeps weak acids largely in their non-ionized form, which favors migration into the organic solvent.
- For basic analytes, the situation is reversed: under acidic conditions, weak bases become protonated and water-soluble, while at higher pH values they remain non-ionized and partition more readily into the organic phase.

This principle underscores how pH serves as a powerful lever in directing analyte behavior during extraction.

Selective Separation: One of the greatest advantages of pH control is its ability to facilitate selective separation of compounds.

- By setting the aqueous pH below the pKa of a weak acid, the analyte remains in its neutral form and is preferentially extracted into the organic solvent. This selective partitioning reduces interference from ionized or polar co-contaminants.
- Similarly, adjusting the pH above the pKa of a weak base ensures the compound remains neutral and favors organic solubility.

Back Extraction Strategies: Back extraction offers an additional refinement step. After an analyte is extracted into the organic solvent, the pH can be reversed in a fresh aqueous phase to re-ionize the compound. This causes the analyte to migrate back into the aqueous layer while leaving neutral impurities behind. Such strategies are particularly valuable in trace-level work, where minimizing background interference is critical for accurate quantification.

Optimization Techniques in pH-Controlled LLE: Several optimization practices ensure reproducibility and efficiency in pH-managed extractions:

- Targeted pH Adjustment: Knowledge of the analyte's pKa is essential. The aqueous phase should be adjusted at least two pH units above the pKa for weak acids or two units below the pKa for weak bases. This guarantees that the analyte is predominantly non-ionized, maximizing transfer into the organic solvent.
- Use of Buffers: Buffer systems, such as phosphate or acetate buffers, stabilize pH throughout the extraction, reducing variability and improving reproducibility.
- Sequential Extractions: Multiple extraction cycles at controlled pH values enhance recovery and purity. For example, an acid initially extracted into the organic phase can undergo back extraction into water at a higher pH for further purification.

- Solvent Selection: The chosen organic solvent must be immiscible with water and demonstrate strong affinity for the non-ionized analyte. Solvent polarity should align with the chemical nature of the compound.
- Ionic Strength Regulation: The addition of salts ("salting out") to the aqueous phase can influence analyte distribution by altering solubility patterns, thereby boosting recovery.
- Temperature Control: Maintaining stable temperature is critical. Fluctuations can alter solubility, partition coefficients, and ultimately the efficiency of extraction.

As shown in Table 6.7, common challenges in pH-controlled LLE such as poor recovery or phase emulsions can be systematically addressed through targeted adjustments.

Key Considerations: Successful pH-controlled LLE hinges on several guiding principles

- Understanding pKa: Adjusting the aqueous phase two units away from the analyte's pKa ensures that the compound is largely non-ionized and therefore more extractable into the organic solvent.
- Ensuring Solvent Immiscibility: Organic solvents must remain immiscible with water to enable phase separation.
- Employing Back Extraction: This step is useful for cleaning up extracts and removing neutral contaminants.
- Maintaining pH Stability: Buffering systems are indispensable for preserving consistent conditions throughout the extraction process.

Table 6.8 outlines a practical guide for aligning extraction conditions with analyte pKa values to maximize selectivity and efficiency. By carefully managing above parameters, LLE can be tailored to deliver high recovery rates, reduced matrix effects, and reproducible results across diverse applications. Forensic toxicology, environmental testing, and anti-doping laboratories alike rely on such optimization strategies to ensure that even low-level analytes can be accurately and reliably detected.

Temperature Management: Temperature plays a pivotal role in shaping the performance of LLE. It influences both the kinetics and equilibrium of the extraction process by modifying solute solubility, solvent viscosity, and diffusivity. Together, these factors determine the partition coefficient the key ratio describing the extent

Table 6.7 Troubleshooting common issues in pH-controlled liquid–liquid extraction

Problem	Solution
Low recovery	Check pH (must be ± 2 units from pKa)
Emulsion formation	Add salt or centrifuge
Poor phase separation	Use less polar solvent (e.g., hexane)
Analyte degradation	Avoid extreme pH (<1 or > 12)

Table 6.8 Optimization framework for LLE based on analyte pKa and aqueous pH

Drug class	Example drugs	Typical pKa	Extraction solvent(s)	Optimal pH (Aqueous Phase)	Remarks
Weak acids	Salicylic acid, Phenobarbital	3–7	Ethyl acetate, diethyl ether, chloroform	pH < pKa (typically 2–3)	Acidic drugs are non-ionized at low pH → extract into organic phase
Weak bases	Amphetamine, Codeine	8–11	Chloroform, DCM, butanol	pH > pKa (typically 9–11)	Basic drugs are non-ionized at high pH → favor extraction into organic
Amphoteric drugs	Ketamine, Levodopa	Multiple (3–9)	Ethyl acetate, isoamyl alcohol	pH ~ 6–8 (tailored by pKa)	Must balance pH to favor non-ionized form depending on desired group
Neutral compounds	Diazepam, Steroids	–	Hexane, ethyl acetate, toluene	pH ~ 7	Not ionizable directly partition into organic solvent
Barbiturates	Phenobarbital, Thiopental	~7.4	Ethyl acetate, DCM	pH 2–4	Weak acids extract at low pH in non-ionized form
NSAIDs	Ibuprofen, Naproxen	~4.5	Diethyl ether, ethyl acetate	pH 2–3	Best recovered at acidic pH
Alkaloids	Atropine, Quinine	7.5–9.5	Chloroform, ether	pH 9–10	Weak bases favor basic pH for non-ionized form
Benzodiazepines	Diazepam, Lorazepam	~3–4 (weak acids)	Ethyl acetate, chloroform	pH 2–4	Prefer acidic pH for extraction
Steroids	Testosterone, Progesterone	–	Hexane, chloroform	Neutral (~7)	Non-ionizable; extract directly
Antidepressants	Amitriptyline, Fluoxetine	~9–10	Chloroform, butanol	pH 10–11	Extracted at basic pH weak bases
Opiates	Morphine, Heroin	~8–9	Chloroform, DCM	pH 9–10	Basic drugs favor high pH extraction

to which an analyte distributes between aqueous and organic phases. Careful regulation of temperature is therefore essential for maximizing extraction efficiency and ensuring reproducible results.

Kinetics and Mass Transfer: Raising the temperature generally reduces solvent viscosity and enhances molecular motion. This acceleration improves diffusion rates and mass transfer between phases, allowing the system to reach equilibrium more quickly. Faster equilibration is advantageous in time sensitive analyses. However, excessive heating may prove counterproductive, as it risks degrading thermo-labile analytes or promoting the evaporation of volatile compounds. Thus, optimization must balance improved kinetics with the preservation of analyte stability.

Equilibrium and Solubility: Temperature shifts can also alter the equilibrium distribution of analytes, depending on whether the extraction is endothermic or exothermic. For many systems, elevated temperatures increase solubility in one phase over another, which may either enhance or hinder recovery. The direction and magnitude of these effects depend on the specific physicochemical properties of the target compound and the solvents used.

Drug-Specific Considerations: The impact of temperature management also varies across drug classes:

- Weak acids and weak bases: Moderate temperature increases often improve extraction efficiency by promoting faster diffusion. However, care must be taken to avoid changes in ionization balance when pH is also a controlled parameter.
- Neutral compounds: Since their extraction is less sensitive to pH, temperature becomes the dominant factor, influencing both phase separation and reproducible partitioning.
- Amphoteric compounds: These analytes contain both acidic and basic functional groups, making their extraction highly dependent on the interplay of temperature and pH. Adjustments must therefore be carefully synchronized to optimize recovery while maintaining chemical stability.

As illustrated in Table 6.9, the influence of temperature on extraction efficiency varies depending on analyte properties, solvent systems, and thermal stability.

Practical Considerations: In most LLE workflows, temperatures are maintained between 20 and 30°C ranging from room temperature to slightly elevated conditions. This moderate range enhances kinetics and phase transfer without significantly increasing the risks of degradation or solvent loss.

The optimal temperature, however, remains highly dependent on the analyte, solvent pair, and biological matrix. For example, volatile compounds require stringent temperature regulation to avoid evaporative losses, while thermally stable drugs may benefit from higher temperatures that shorten extraction time.

Another important aspect is the integration of temperature control with pH optimization. For analytes that are pH-sensitive, such as weak acids or bases, temperature must be adjusted without disturbing ionization equilibria. A fine balance

Table 6.9 Effects of temperature on LLE for selected drug classes

Drug class	Example drug	Typical temperature range (°C)	Common extraction solvent(s)	Key notes/effect of temperature
Weak acids	Ibuprofen	20–30	Ethyl acetate, diethyl ether	At ~25 °C, enhanced diffusion improves extraction; moderate heating reduces viscosity, favoring analyte transfer
Weak bases	Amphetamine	20–30	Chloroform, dichloromethane (DCM)	Extraction at room temperature benefits from increased diffusion without significant evaporation risk; slight elevation can improve kinetics
Neutral compounds	Diazepam	25–30	Hexane, toluene, ethyl acetate	Generally, gentle heating (up to 30°C) increases mass transfer rates while maintaining compound stability
Amphoteric drugs	Tetracycline	20–30	Ethyl acetate, isoamyl alcohol	Temperature must be optimized alongside pH adjustments to ensure both ionizable groups are in the desired form
Steroids	Testosterone	20–30	Chloroform, dichloromethane	Although non-ionizable, extraction benefits from moderate temperatures to speed up equilibrium achievement without degradation

between these two parameters ensures that the partition coefficient is maximized while safeguarding analyte integrity.

Temperature is not an isolated factor but one that interacts with pH control, solvent polarity, and ionic strength to define the overall success of LLE. By tailoring temperature conditions to the unique requirements of each analyte class, analysts can achieve efficient, selective, and reproducible extractions across a wide range of forensic, clinical, and anti-doping applications.

Multiple Extractions: In LLE, carrying out several extractions with smaller portions of solvent is often more effective than performing a single extraction with a larger volume. This approach takes advantage of the principles of partitioning, allowing analytes to be transferred more completely into the organic phase across multiple steps.

Principles of Multiple Extractions: The efficiency of LLE is governed by the partition coefficient (KD), which defines the relative distribution of a solute between aqueous and organic phases. When multiple extractions are performed, each successive step lowers the concentration of the solute remaining in the original phase, resulting in a more complete transfer into the extracting solvent. This strategy enhances analyte recovery and reduces the risk of losses that can occur in one-step procedures.

Methodological Steps: The process begins with thorough mixing of the aqueous and organic phases, typically in a separatory funnel, to allow solute partitioning. After complete phase separation, the organic layer is collected, while the aqueous layer undergoes further extractions with fresh portions of solvent. The organic extracts from each step are then pooled and concentrated, producing a purified sample ready for analysis.

Practical Considerations: Effective multiple extractions rely on careful solvent selection, ensuring that the chosen solvent has a strong affinity for the non-ionized form of the analyte. Attention must also be given to preventing emulsion formation, as emulsions can impede phase separation and reduce efficiency.

- Mixing phases: Combine the aqueous and organic phases in a separatory funnel, ensuring thorough mixing to allow the solute to partition between the two layers.
- Phase separation: Allow the layers to separate completely. The denser liquid will form the bottom layer, while the less dense liquid will be on top.
- Extraction repetition: Remove the organic layer containing the extracted solute. Repeat the extraction process with fresh portions of the organic solvent to further reduce the solute concentration in the aqueous phase.
- Combining extracts: Combine the organic extracts from each step. This pooled extract can then be processed further, such as by evaporating the solvent to isolate the desired compound.

Advantages

- Improved recovery: More solute is transferred into the organic phase across multiple steps.
- Efficient solvent use: Smaller portions across several extractions are often more economical than a single large volume.
- Enhanced purity: Repeated partitioning reduces co-extracted impurities, improving the clarity of the final extract.

By combining this technique with optimized pH, solvent choice, and temperature control, multiple extractions provide a powerful way to increase both efficiency and reproducibility in forensic and anti-doping applications.

6.2.3 Workflow of LLE

The process of LLE follows a series of well-defined steps designed to efficiently isolate analytes from complex biological matrices. As illustrated in Fig. 6.2, the LLE workflow involves sequential steps beginning with sample preparation, followed by

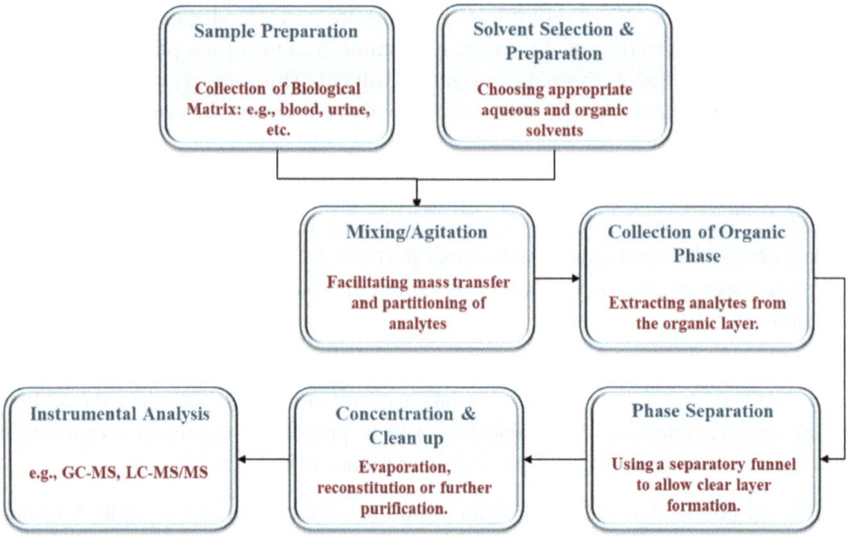

Fig. 6.2 Flowchart illustrating the sequential steps of LLE

solvent mixing, phase separation, optional washing, evaporation and reconstitution, and final analysis.

Step 1. Sample Preparation: The first stage involves conditioning the sample to ensure optimal analyte recovery. The aqueous phase pH is adjusted according to the chemical nature of the analyte for example, raising the pH to around 9 facilitates extraction of basic drugs in their uncharged form. Salts such as sodium chloride may be added to promote the salting-out effect, thereby reducing analyte solubility in water and encouraging migration into the organic solvent. Viscous samples, such as blood, are often diluted to improve handling and enhance solvent sample interactions.

Step 2. Mixing with Solvent: An immiscible organic solvent commonly ethyl acetate or dichloromethane is introduced. The mixture is vigorously shaken or vortexed to maximize contact between the two phases, allowing analytes to partition effectively according to their polarity.

Step 3. Phase Separation: The system is then allowed to separate, either by gravity in a separatory funnel or through centrifugation. The phase containing the analyte is carefully collected: non-polar compounds typically migrate into the organic layer, while highly polar compounds may remain in the aqueous layer.

Step 4. Washing (Optional): To enhance purity, the collected extract may undergo a washing step using a mild aqueous buffer. This back-extraction helps remove residual proteins or polar interferences.

Step 5. Evaporation and Reconstitution: The organic solvent is evaporated, often under a gentle nitrogen stream. The dried residue is reconstituted in a solvent compatible with the intended analytical technique, such as methanol for LC–MS applications.

Step 6. Analysis: The prepared extract is finally analyzed using appropriate instrumentation: LC–MS for polar, non-volatile analytes or GC–MS for volatile compounds.

6.2.4 Advantages of LLE

LLE remains one of the most widely used techniques in forensic toxicology and anti-doping laboratories due to its balance of simplicity, flexibility, and efficiency.

- **Simplicity and Accessibility**: One of the greatest strengths of LLE lies in its straightforward design. The method requires no advanced instrumentation, relying instead on readily available laboratory glassware such as separatory funnels and standard organic solvents. This ease of implementation makes LLE especially valuable in forensic settings where rapid sample preparation is often essential.
- **Cost Effectiveness**: LLE is an economical approach compared to many modern extraction methods. The solvents and equipment required are relatively inexpensive, making the technique highly appealing for laboratories that must balance analytical rigor with financial limitations. As a result, it continues to serve as a preferred method for routine analyses in resource-conscious environments.
- **Versatility**: Another major advantage is the adaptability of LLE. By altering parameters such as solvent choice, pH adjustment, and extraction time, the method can be tailored to recover a wide spectrum of analytes including drugs of abuse, therapeutic agents, metabolites, and toxicants from diverse biological matrices such as blood, urine, and hair. This versatility establishes LLE as a reliable first-line screening tool.
- **Selectivity for Non-polar Compounds**: Because most organic solvents are non-polar, LLE is particularly effective for isolating neutral and moderately polar compounds. When the physicochemical properties of target analytes are considered during method optimization, LLE can achieve strong recoveries and reproducible results.
- **Scalability**: Finally, LLE offers flexibility in terms of sample size. It can be applied to both small and large sample volumes, and sequential extractions can be incorporated to maximize analyte recovery. This scalability makes it highly practical for the variable sample quantities encountered in forensic casework.

Together, these advantages explain why LLE remains a cornerstone in toxicological and anti-doping analyses despite the rise of more advanced extraction technologies.

6.2.5 Disadvantages of LLE

While LLE remains a widely applied technique in forensic toxicology and anti-doping studies, it is not without notable limitations.

- High Solvent Demand and Environmental Concerns: A major drawback of LLE is the large volume of organic solvents required. This not only raises operational costs but also increases the burden of hazardous waste management. Safe disposal practices are essential, and strict laboratory safety protocols must be followed to minimize environmental and occupational risks.
- Time and Labor Intensive: The workflow of LLE often involves multiple extraction cycles to achieve sufficient analyte recovery. Each cycle requires mixing, separation, and collection of the desired phase, making the process both time-consuming and labor-intensive. Without standardized procedures, variability between operators can further compromise reproducibility.
- Emulsion Formation: One persistent challenge is the formation of emulsions during mixing. These stable mixtures hinder effective phase separation and may result in poor analyte recovery or even sample loss. Emulsions are especially problematic in biological samples rich in proteins or lipids.
- Limited Automation: Compared to more modern techniques like SPE, LLE is less adaptable to automated systems. This reduces throughput in high-volume laboratories and increases dependence on manual handling, potentially introducing inconsistencies.
- Sensitivity to Conditions: The efficiency of LLE is highly dependent on precise control of pH and temperature. Small deviations can alter analyte partitioning and reduce recovery. Moreover, LLE is poorly suited for highly polar compounds, which often remain in the aqueous phase, necessitating alternative extraction approaches.

6.2.6 Summary

LLE remains a valuable tool in forensic toxicology, particularly as an accessible and cost-effective approach for the preliminary isolation of diverse analytes. Its appeal lies in its straightforward design, scalability across different sample sizes, and adaptability to a range of compounds, with notable effectiveness for non-polar substances. Despite these strengths, LLE is not without challenges. High solvent consumption raises both economic and environmental concerns, while the risk of emulsion formation and the reliance on time-intensive manual handling can compromise efficiency and reproducibility. Furthermore, limited suitability for automation restricts its utility in high-throughput laboratory settings. In forensic investigations, where accuracy and consistency are paramount, such limitations often necessitate the use of complementary or alternative extraction methods. Ultimately, careful consideration of both its advantages and drawbacks allows practitioners to determine when LLE is appropriate and when more advanced techniques may deliver superior performance.

6.3 Solid-Phase Extraction (SPE)

SPE is a widely employed sample preparation technique in forensic toxicology, designed to isolate, purify, and concentrate analytes from complex biological matrices such as blood, urine, hair, and oral fluids [28, 29]. The primary objective of SPE is to simplify the sample by removing interfering components, thereby improving analyte recovery and detection. This approach significantly enhances the sensitivity and reproducibility of advanced analytical platforms, particularly LC–MS/MS and GC–MS. By combining selectivity, efficiency, and adaptability, SPE offers a robust method for handling diverse analytes and matrices in forensic and anti-doping contexts. Its structured workflow not only improves analytical clarity but also reduces matrix effects, making it a cornerstone of modern toxicological investigations.

6.3.1 Theory and Principle of SPE

SPE is based on the principle of selective adsorption and partitioning [30–32]. The core process involves the following stages:

- **Solid Sorbent Selection**: SPE utilizes cartridges or columns packed with solid adsorbents such as C18, silica, polymeric phases, or ion-exchange resins. The choice of sorbent is critical, as each material interacts differently with target analytes depending on their polarity, charge, or structural characteristics.
- **Adsorption**: When the biological sample is passed through the cartridge, analytes of interest interact with the solid phase via hydrophobic forces, hydrogen bonding, ionic attraction, or mixed-mode mechanisms. This step selectively retains analytes while minimizing non-specific binding.
- **Washing**: A carefully chosen solvent is applied to remove weakly bound or irrelevant matrix components. This stage enhances selectivity by eliminating potential interferences without dislodging the analytes.
- **Elution**: Finally, analytes are released from the sorbent using solvents or solvent mixtures that disrupt their binding interactions. The result is a purified, concentrated extract that is ready for precise instrumental analysis.

As illustrated in Fig. 6.3, the SPE workflow involves sequential steps of adsorption, washing, and elution, enabling the selective isolation of target analytes from complex biological samples.

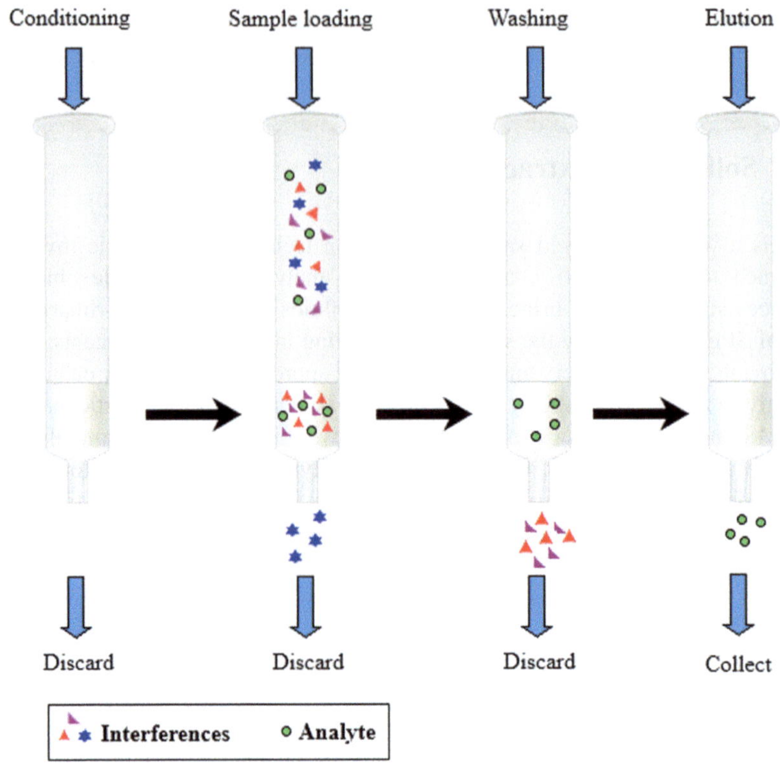

Fig. 6.3 Schematic representation of the SPE process

6.3.2 SPE Systems: Types and Comparison

SPE has become one of the most widely adopted approaches for sample preparation in forensic, clinical, pharmaceutical, and environmental laboratories. The technique is based on the principle that analytes of interest can be selectively retained on a sorbent bed, washed to remove impurities, and then eluted in purified form for downstream analysis. A typical SPE workflow involves four core steps: conditioning the sorbent to activate binding sites, loading the sample to allow interaction between analytes and the sorbent, washing to remove unwanted matrix components, and elution to release purified analytes.

Although the principle is universal, SPE can be performed using a variety of devices that differ in the way fluids are passed through the sorbent bed. The driving force can be created by vacuum (negative pressure), compressed gas (positive pressure), centrifugation, or fully automated robotic systems. Each system has its own operational strengths, limitations, and areas of application, making it important to compare them when selecting the most appropriate platform for forensic and toxicological workflows.

Vacuum (Negative-Pressure) SPE: The vacuum manifold remains one of the most traditional and accessible SPE systems. In this setup, cartridges or 96-well SPE plates are arranged in a sealed manifold chamber connected to a vacuum pump. When negative pressure is applied, liquid is drawn through the sorbent bed into a collection vessel. The glass-block manifold design often allows for clear visual inspection of flow, making it a practical tool for many laboratories. As shown in Fig. 6.4, the vacuum applies negative pressure to pull liquid through the sorbent bed, enabling simultaneous processing of multiple samples.

Advantages: Vacuum SPE is simple, cost-effective, and suitable for a broad range of applications. It supports diverse sample volumes, from small biological samples to larger environmental extracts. Multi-position manifolds can accommodate 20–36 cartridges simultaneously, offering moderate throughput. Because it does not require a compressed gas supply, the infrastructure demands are minimal only a vacuum pump and waste collection system are necessary.

Disadvantages: The major limitation of vacuum SPE is uneven flow control. As some wells or cartridges empty faster than others, the vacuum level rises, leading to variability in flow rates. This inconsistency can cause sorbent drying, incomplete recoveries, or poor reproducibility across replicates. Additionally, unused ports must be carefully blocked, and strong pumps with airtight seals are needed to maintain performance. Variability in operator handling can further add to irreproducibility.

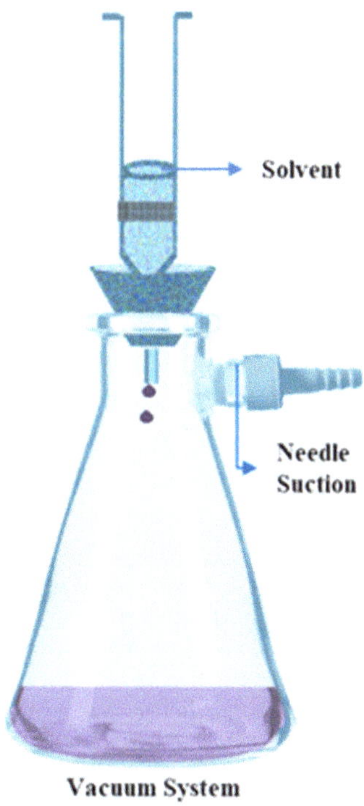

Fig. 6.4 Schematic representation of vacuum (negative-pressure) SPE

Applications: Vacuum SPE is commonly used in environmental monitoring (e.g., concentrating pesticides or pollutants from water) and in toxicology for drug extraction from blood or urine. It is cost-efficient for general cleanup prior to GC, LC, or mass spectrometric analysis. For laboratories on a budget, vacuum manifolds remain reliable workhorses.

Positive-Pressure SPE: Positive-pressure systems employ compressed gas usually clean, inert nitrogen to push liquids through the sorbent bed from above. In these manifolds, cartridges or plates are sealed, and a regulated gas supply provides uniform pressure across all channels. Flow control valves and gauges allow precise adjustment, ensuring reproducibility even when some wells become empty. As illustrated in Fig. 6.5, compressed nitrogen applies uniform downward pressure across all cartridges or wells, ensuring consistent and reproducible extraction flow rates.

Advantages: Positive-pressure systems produce highly uniform and reproducible results compared to vacuum systems. Because each well experiences the same regulated pressure, recoveries are more consistent, and the risk of sorbent drying is minimized. This setup is particularly beneficial for viscous or particle-rich samples, such as plasma, serum, or sludge, where vacuum manifolds struggle. Inert nitrogen prevents contamination, and advanced devices often allow independent control of each well, providing high precision.

Disadvantages: The need for a compressed-gas supply and specialized hardware makes positive-pressure SPE more costly. Systems require regulators, pneumatic

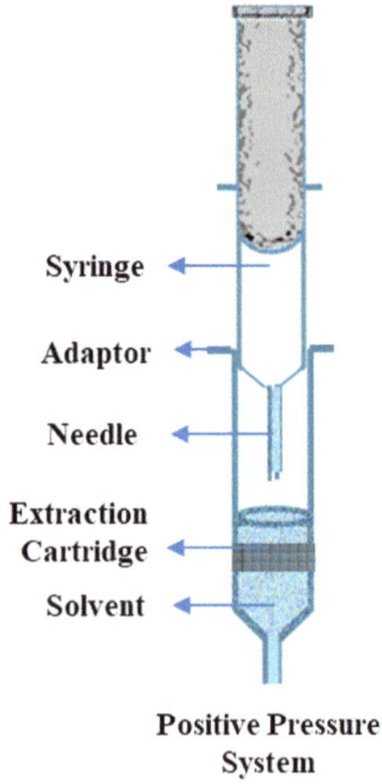

Fig. 6.5 Schematic representation of positive-pressure SPE

seals, and compatible cartridges or plates, adding to the complexity. Smaller laboratories may find these investments difficult to justify, especially if throughput needs are modest. There is also a risk of splashing or aerosolization if pressure is not properly controlled.

Applications: Positive-pressure systems are widely used in regulated and high-throughput environments, including pharmaceutical bioanalysis, forensic toxicology, and clinical testing. They are especially suited for LC–MS/MS workflows, where precision and reproducibility are critical. Systems like Waters Positive Pressure-96 or PCI's SPE-96 can process 48–96 samples simultaneously, making them ideal for large-scale laboratories.

Centrifugal (Plate-Based) SPE: Centrifugal SPE makes use of a centrifuge to drive solvents through sorbent media, usually arranged in 96-well microplate formats. A sorbent-packed plate is placed above a collection plate, and centrifugation at forces of 100–400 g pushes liquids through the sorbent. Systems such as Empore 96-well plates are designed for compatibility with centrifuges, vacuum, or pressure-based setups. As depicted in Fig. 6.6, centrifugal force drives solvents through the sorbent beds of a 96-well SPE plate, enabling parallel processing of multiple samples.

Advantages: Centrifugal SPE is simple to implement, requiring only a standard laboratory centrifuge with a plate rotor. All wells are subjected to identical force, ensuring uniform flow across the plate. This setup is well-suited for small volumes and microelution formats, which allow highly concentrated extracts with minimal solvent evaporation. It also offers high throughput, enabling the simultaneous processing of 96 samples with little hands-on time. When plates are properly sealed, the risk of cross-contamination is minimal, and flow rates can be adjusted easily by changing centrifugation speed and duration.

Disadvantages: Centrifugal SPE is limited to microplate formats and is unsuitable for large-volume cartridges. The small sample capacity (often ≤2 mL per well) restricts

Fig. 6.6 Schematic representation of centrifugal (plate-based) SPE

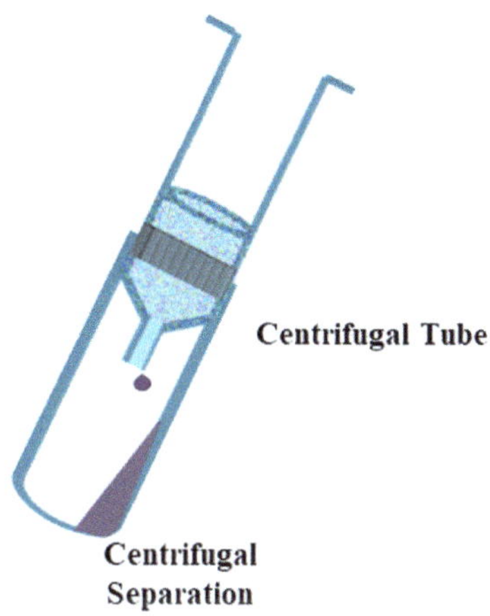

its use to applications requiring small sample volumes. Additionally, the reliance on specialized, and often proprietary, microplates increases consumable costs. Careful optimization of centrifugation parameters is essential to avoid incomplete extraction or analyte breakthrough.

Applications: This method is popular in high-throughput screening, pharmacokinetics, and bioanalytical workflows, particularly when handling large numbers of small-volume samples. It is also used in proteomics and genomics for buffer exchange or desalting. Because the plates follow standard microplate dimensions, centrifugal SPE integrates well with automated liquid-handling systems.

Automated/Robotic SPE Systems: Automated SPE platforms represent the most advanced evolution of this technique. These systems integrate the conditioning, loading, washing, and elution steps into robotic workflows controlled by specialized software. Devices include dedicated benchtop SPE workstations (e.g., ThermoFisher Dionex AutoTrace, Gilson ASPEC, Biotage Extrahera) and modular liquid-handling robots with SPE capabilities (e.g., Hamilton MPE^2, Tecan Resolvex, Agilent Bravo + PPM).

Advantages: Automation delivers unmatched throughput, reproducibility, and safety. Robots can handle dozens to hundreds of samples with minimal supervision, significantly reducing labor. By maintaining constant flow rates and regulated pressures, automated SPE minimizes variability between runs. Built-in safeguards such as pressure sensors prevent clogging or overpressure. Enclosed systems also protect operators from solvent exposure. Additionally, robotic systems can integrate complementary steps such as evaporation, dilution, and filtration, creating end-to-end sample preparation workflows.

Disadvantages: The primary drawbacks are cost, complexity, and maintenance. Automated SPE units are expensive and may require specialized training for operation. For laboratories processing only small batches, the investment may not be justified. Maintenance downtime and the need for method validation add further considerations in regulated environments.

Applications: Automated SPE is ideal for high-throughput laboratories that demand consistent and reliable results, such as pharmaceutical R&D, clinical testing under GLP/GMP, and environmental monitoring. Forensic and anti-doping laboratories also benefit from automation, especially when reproducibility and scalability are essential. Comparative studies have shown automated positive-pressure systems achieving nearly complete recoveries with lower variability compared to manual vacuum-based systems.

6.3.3 *Overview of SPE Cartridges*

SPE remains an essential tool in forensic toxicology because it enables selective cleanup and pre-concentration of analytes from complex biological samples. A key determinant of extraction success is the proper selection of the SPE cartridge, which must be carefully matched to both the physicochemical properties of the analyte and the complexity of the matrix. By understanding the range of cartridge chemistries available from conventional reversed-phase to more advanced mixed-mode and molecularly imprinted polymers laboratories can tailor their methods to achieve higher sensitivity, selectivity, and reproducibility. As illustrated in Fig. 6.7, SPE cartridges consist of sorbent-packed columns designed to selectively retain analytes while allowing interferences to be washed away before elution.

SPE cartridges are produced in various formats to address different analytical needs. In forensic applications, they are widely employed to recover drugs of abuse, therapeutic agents, metabolites, and toxins from matrices such as blood, urine, hair, and oral fluid.

Several factors guide the selection of a suitable cartridge:

- **Nature of the analyte**: Acidic, basic, neutral, or amphoteric properties influence the type of sorbent required.
- **Matrix complexity**: Highly complex samples demand stronger cleanup capacity to minimize interferences.
- **Solvent compatibility**: Cartridges must tolerate the solvents chosen for conditioning, washing, and elution steps.
- **Retention mechanism**: Options include hydrophobic binding (reversed-phase), ionic interactions (ion-exchange), or combinations of both in mixed-mode cartridges.

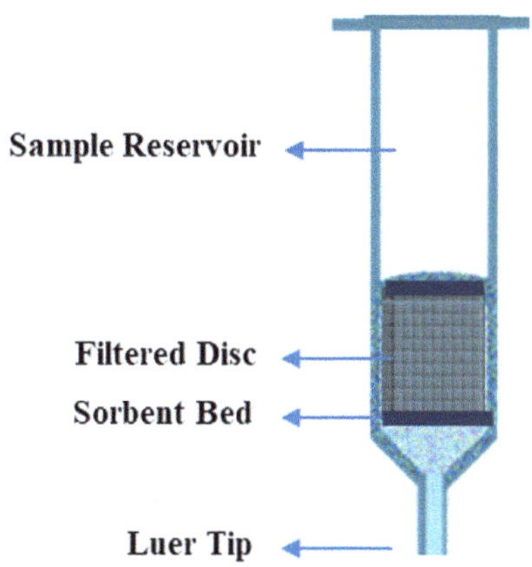

Fig. 6.7 Schematic representation of SPE cartridge

Careful alignment of cartridge chemistry with analyte and matrix characteristics ensures efficient sample preparation, thereby improving downstream chromatographic and mass spectrometric analysis.

6.3.4 SPE Sorbent Materials: Composition, Properties, and Applications

SPE has become one of the most widely used sample preparation strategies in forensic toxicology, clinical chemistry, and environmental analysis. Its efficiency depends largely on the type of sorbent employed, since sorbent chemistry governs analyte retention, selectivity, and elution behavior. Broadly, SPE sorbents can be grouped into silica-based sorbents, polymeric sorbents, ion-exchange phases, and mixed-mode systems, each offering distinct retention mechanisms and operational characteristics. Understanding their properties is essential for selecting the right sorbent for specific applications.

Silica-Based Sorbents: Silica-based sorbents are among the earliest and most widely used in SPE. These sorbents are composed of porous silicon dioxide (SiO_2) particles, which typically exhibit very high surface areas, often ranging from 500 to 800 m^2/g. Their surfaces are covered with abundant silanol (Si–OH) groups, which impart polarity and enable hydrogen bonding, dipole–dipole interactions, and ion–dipole forces with analytes.

The raw silica particles used in SPE are usually 40–60 μm in diameter, with pore sizes in the range of 60–100 Å. Unmodified silica, also referred to as bare silica, is commonly used in normal-phase SPE to retain highly polar compounds. However, to extend silica's application to non-polar analytes, chemical modification of the surface is performed. For instance, octadecyl (C18) or octyl (C8) chains are covalently bonded to silanol groups, creating reversed-phase sorbents that retain analytes through hydrophobic interactions.

C18 sorbents provide strong retention for non-polar and moderately polar compounds, while C8 phases offer milder hydrophobic selectivity, which can be useful when shorter elution times or weaker retention are desired. Residual silanol groups that remain after bonding can contribute to unwanted interactions, particularly with basic analytes. To reduce this effect, "end-capping" techniques are employed, where trimethylsilyl groups react with free silanols, minimizing their contribution to secondary interactions.

Despite their versatility, silica-based sorbents have certain limitations. They are restricted to a pH operating window of approximately 2–9. Beyond this range, silica can dissolve or lose functionality. Furthermore, silanol groups become ionized above pH ~ 4.5, potentially introducing unwanted cation-exchange interactions that complicate the retention of basic compounds. Nevertheless, silica-based C18 and C8 sorbents remain widely used for extracting pharmaceuticals, metabolites, and organic pollutants from aqueous samples due to their high reproducibility and efficiency.

Polymeric Sorbents: Polymeric sorbents were developed to overcome some of the limitations of silica, particularly its restricted pH range and surface chemistry. These sorbents are typically made from cross-linked organic polymers, most commonly polystyrene–divinylbenzene (PS–DVB). Produced as spherical beads, polymeric sorbents offer high surface areas, often exceeding 1000 m^2/g, and uniform particle structures that allow for consistent packing and reduced backpressure in SPE cartridges.

Unlike silica, polymeric sorbents do not contain silanol groups, which means they are chemically stable across the full pH range (pH 1–14). This enables their use with harsh solvents and extreme conditions without risking sorbent degradation. One of the most widely adopted designs is the hydrophilic–lipophilic balanced (HLB) polymer, such as the Oasis HLB sorbent. These materials combine hydrophobic DVB with a polar monomer, such as N-vinylpyrrolidone, yielding dual retention capabilities. This balance allows simultaneous retention of polar, non-polar, acidic, and basic analytes, making HLB phases versatile for multi-analyte applications.

Other polymeric sorbents incorporate polar functional groups like methacrylate or amides into the DVB backbone, enhancing interactions with specific classes of compounds. The presence of aromatic DVB units also introduces π–π interactions, beneficial for analytes containing aromatic rings. Another special category of polymeric sorbents includes graphitized carbon black, which strongly retains highly polar molecules such as small acids, nucleotides, or polar metabolites, often overlooked by conventional reversed-phase sorbents.

Polymeric sorbents are thus particularly valuable in forensic toxicology, where complex biological matrices demand robustness, broad analyte coverage, and reproducibility.

Ion-Exchange Sorbents: While silica and polymeric sorbents rely mainly on hydrophobic or polar interactions, ion-exchange sorbents exploit electrostatic forces between charged analytes and surface-bound functional groups. These phases are designed with covalently attached ionic groups that attract oppositely charged molecules.

Ion-exchange sorbents can be divided into cation exchangers and anion exchangers, each of which may be classified as strong or weak depending on their ionization behavior across the pH scale:

- *Strong Cation Exchange (SCX)*: Contains sulfonic acid ($–SO_3H$) groups, which remain ionized ($–SO_3^-$) across all pH values. SCX sorbents are ideal for binding protonated basic drugs.
- *Weak Cation Exchange (WCX)*: Uses carboxyl groups (–COOH), which ionize to $–COO^-$ only above pH 4–5. WCX phases offer selective retention of cations in moderately basic environments.
- *Strong Anion Exchange (SAX)*: Contains quaternary ammonium ($–N^+(CH_3)_3$) groups, permanently charged and thus capable of binding acidic, negatively charged analytes across all pH values.
- *Weak Anion Exchange (WAX)*: Employs amine groups ($–NH_2$/–NHR), protonated at low pH, to bind acidic analytes selectively.

These functional groups may be grafted onto silica or polymeric supports, combining the structural benefits of the base material with ionic selectivity. Typical workflows involve adjusting the sample pH to ensure analytes are in the appropriate charged form for retention, followed by elution through pH changes or by introducing counter-ions.

Ion-exchange SPE is particularly effective for concentrating charged drugs, pesticides, and organic acids from biological and environmental samples. Forensic toxicology often employs SCX sorbents to isolate basic drugs such as amphetamines or opioids from plasma, while SAX sorbents may be used for acidic metabolites or environmental pollutants.

Mixed-Mode Sorbents: Mixed-mode sorbents represent an advanced category that integrates two retention mechanisms typically hydrophobic and ionic on the same support. This dual functionality enhances selectivity, as analytes must interact through both mechanisms to be retained.

For example, a C8-SCX sorbent combines octyl hydrophobic groups with sulfonic acid moieties. A basic analyte that is protonated under the chosen pH conditions can simultaneously partition into the hydrophobic layer and bind electrostatically to the sulfonate groups. Similarly, C8-SAX phases incorporate quaternary ammonium groups alongside hydrophobic chains, enabling strong retention of acidic analytes through combined hydrophobic and ionic interactions.

The strength of mixed-mode SPE lies in its ability to discriminate between target analytes and interferences that lack one or both types of interaction. By carefully tuning the loading solvent, washing conditions, and elution strategy, highly selective cleanup can be achieved, even from very complex matrices such as whole blood or postmortem tissues.

Applications in forensic toxicology include the extraction of drugs of abuse, therapeutic drugs, and metabolites from biological samples where traditional single-mode phases may fail to deliver adequate purity. For instance, mixed-mode sorbents are particularly advantageous when analytes must be isolated from a background of structurally similar compounds or endogenous interferences.

Practical Considerations in Sorbent Selection: The choice of sorbent depends on several factors:

- *Chemical nature of the analyte*: Neutral, acidic, basic, amphoteric, or highly polar.
- *Matrix complexity*: Simple aqueous solutions versus complex biological matrices.
- *Required selectivity and sensitivity*: Routine screening versus confirmatory analysis.
- *Operational conditions*: pH range, solvent compatibility, and robustness required.

For general non-polar analytes, silica-based C18 or polymeric HLB sorbents are reliable. For highly polar molecules, graphitized carbon or polar-modified sorbents are more suitable. Charged analytes often require ion-exchange sorbents, while complex forensic samples benefit most from mixed-mode cartridges that combine hydrophobic and ionic retention. An overview of commonly used SPE sorbent materials, including their mechanisms, applications, and practical advantages and limitations, is summarized in Table 6.10. The principal SPE parameters relevant to different drug classes are summarized in Table 6.11.

6.3.5 Types of SPE Cartridges

SPE relies heavily on the type of cartridge selected, as this determines the specificity, recovery, and overall success of the analytical process. In forensic toxicology, where accuracy and selectivity are paramount, cartridges are designed with distinct sorbent chemistries to match the chemical properties of analytes and the complexity of biological matrices. Below is an overview of the most commonly used SPE cartridge types, their materials, solvent compatibility, and applications in drug analysis.

Reversed-Phase SPE Cartridges: Reversed-phase cartridges are among the most widely used in forensic toxicology. These typically contain non-polar sorbents such as silica-based or polymeric materials bonded with hydrophobic groups, most commonly C18 (octadecyl) or C8 (octyl). The retention mechanism relies on hydrophobic interactions, which allow neutral and moderately polar compounds to bind effectively to the sorbent surface.

These cartridges are highly compatible with organic solvents such as methanol and acetonitrile, either alone or mixed with water, which makes them flexible for both washing and elution steps. Reversed-phase cartridges are particularly useful for analytes such as benzodiazepines, cannabinoids, opiates, and several metabolites. For instance, a C18 cartridge is often employed to extract diazepam from serum, where the drug binds to the hydrophobic surface and is later released with a strong organic solvent.

Ion-Exchange SPE Cartridges: Ion-exchange cartridges introduce an additional layer of selectivity by incorporating functional groups capable of electrostatic interactions. These sorbents may be silica- or polymer-based and are functionalized with groups such as sulfonic acids (for strong cation exchange) or quaternary amines (for strong anion exchange).

Solvent compatibility is generally robust in aqueous and moderately organic systems, but maintaining the correct pH is essential to ensure ionization of both the analyte and the sorbent. Cation-exchange cartridges are commonly used for basic drugs such as amphetamines and opioids, provided the sample pH is adjusted so the drugs are positively charged. Conversely, anion-exchange cartridges target acidic drugs such as NSAIDs, where analytes remain negatively charged under acidic conditions. This dual flexibility makes ion-exchange cartridges indispensable in routine forensic applications.

Mixed-Mode SPE Cartridges: Mixed-mode cartridges combine hydrophobic (reversed-phase) and ionic (ion-exchange) retention mechanisms into a single

Table 6.10 Common SPE sorbent materials: examples, retention mechanisms, applications, advantages, and limitations

Type	Examples	Mechanism	Applications	Advantages	Limitations
Reversed-phase	C18, C8, C2, Phenyl groups	Hydrophobic interactions between non-polar analytes and alkyl-bonded silica	Extraction of non-polar compounds (e.g., drugs, PAHs) from polar matrices (e.g., blood, water)	High retention of hydrophobic analytes; versatile for organic solvents	Poor retention of polar compounds; requires organic solvents for elution
Normal-phase	Silica, Alumina, Florisil (magnesium silicate)	Polar interactions (e.g., hydrogen bonding, dipole–dipole) with polar analytes	Isolation of polar compounds (e.g., carbohydrates, steroids) from non-polar matrices	Effective for hydrophilic analytes; compatible with non-polar eluents	Sensitive to moisture; requires anhydrous conditions
Ion-exchange	SAX (Strong Anion Exchange), SCX (Strong Cation Exchange),WCX, WAX	Electrostatic interactions between charged analytes and oppositely charged sorbents	Purification of charged molecules (e.g., proteins, organic acids, nucleotides)	High selectivity for ionic species; adjustable via pH and ionic strength	Requires sample pre-treatment (pH adjustment); limited to ionizable compounds
Mixed-mode	Oasis HLB, Strata-X-CW, Bond Elut PLEXA	Combines reversed-phase and ion-exchange mechanisms for dual selectivity	Extraction of analytes with mixed polarity/ charge (e.g., opioids, amphetamines)	Broad applicability; reduces matrix interference in complex samples	Higher cost; requires optimized elution protocols
Molecularly imprinted polymers (MIPs)	Custom-synthesized polymers (e.g., for caffeine, pesticides)	Template-specific cavities mimic antibody–antigen binding for target analytes	Selective extraction of biomarkers, toxins, or drugs in biological/ environmental samples	Exceptional specificity; reusable	Time-consuming synthesis; limited to pre-defined targets

(continued)

Table 6.10 (continued)

Type	Examples	Mechanism	Applications	Advantages	Limitations
Hydrophilic–lipophilic balance (HLB)	Oasis HLB, Plexa PCX	Balanced polarity for simultaneous retention of hydrophilic and hydrophobic analytes	Multi-residue analysis in environmental samples (e.g., pesticides, pharmaceuticals)	High capacity; no conditioning required for aqueous samples	Limited retention for extremely polar or non-polar compounds
Carbon-based	Graphitized Carbon Black (GCB), Activated Carbon	π-π interactions, van der Waals forces, and polar functional group adsorption	Extraction of polar compounds (e.g., phenols, dyes) and removal of pigments	Effective for polar analytes; removes matrix interferents (e.g., chlorophyll)	Irreversible adsorption of some analytes; requires strong eluents (e.g., TFA)
Immuno-affinity	Antibody-coated sorbents (e.g., for aflatoxins, hormones)	Antigen–antibody binding for highly specific extraction	Targeted analysis of low-abundance biomarkers (e.g., hormones, mycotoxins)	Unmatched specificity; ultra-low detection limits	High cost; antibodies degrade over time

Table 6.11 Key Parameters for SPE across different drug classes

Drug class	Example drugs	Sorbent type	Optimal pH	Elution solvent/system
Basic drugs	Amphetamines, Codeine	Mixed-mode Cation Exchange; C18	pH 9–11	Organic solvent with acidic modifier (e.g., methanol + 0.1% formic acid)
Acidic drugs	Ibuprofen, Naproxen	Reverse-phase (C18); Anion Exchange	pH 2–3	Solvent with increased organic content (e.g., acetonitrile)
Neutral drugs	Diazepam, Benzodiazepines	C18 reverse-phase	Neutral (pH 6–7)	High percentage organic solvents (e.g., methanol, acetonitrile)
Amphoteric drugs	Tetracycline, Levofloxacin	Mixed-mode (Cation/Anion Exchange or C18)	Depends on pKa; typically pH 6–8	Tailored solvent system combining aqueous buffer and organic modifiers

sorbent. These are usually polymer-based materials with dual functional groups, such as a C18 alkyl chain combined with strong or weak ion-exchange sites.

The strength of mixed-mode cartridges lies in their versatility. They perform effectively in both aqueous and organic solvent systems and can be applied across a wide pH range. Such cartridges are especially valuable when dealing with complex biological matrices or analytes that have both hydrophobic and ionic properties. Antidepressants, for example, often require both hydrophobic and ionic interactions for effective extraction, making mixed-mode cartridges the preferred choice.

Hydrophilic–Lipophilic Balanced (HLB) Cartridges: Hydrophilic–lipophilic balanced (HLB) cartridges represent a more recent innovation in SPE technology. These cartridges are made from polymeric copolymers designed to provide balanced interactions for both polar and non-polar analytes. By combining hydrophobic and hydrophilic functionalities within a single sorbent, HLB cartridges offer broad versatility and strong retention across a wide polarity range.

They are chemically stable in water-miscible solvents such as methanol and acetonitrile and can withstand a wide pH spectrum, making them especially robust for forensic applications. The Oasis HLB cartridge is a widely used option, often applied in multi-residue drug testing. This type of cartridge is particularly effective in handling complex samples, ensuring reliable recovery and cleanup for a broad array of drugs and metabolites. Guidelines for choosing appropriate SPE cartridges for forensic toxicology are provided in Table 6.12.

Molecularly Imprinted Polymers (MIPs): Molecularly imprinted polymer (MIP) cartridges offer the highest degree of selectivity among SPE formats. These cartridges

Table 6.12 Selection guide for SPE cartridges in forensic toxicology applications

Cartridge type	Packing material	Drug class	Optimal pH	Elution solvent
C18 (RP)	Octadecyl silica	Opioids, cannabinoids	2–7	Methanol/ethyl acetate
SCX (IE)	Sulfonic acid	Amphetamines, PCP	<pKa	NH_4OH/ methanol
MCX (Mixed)	C18 + SCX	Broad spectrum	2–7	2% NH_4OH in methanol
Silica (NP)	Si–OH	LSD, polar drugs	Neutral	Methanol

are engineered with synthetic polymers containing "molecularly imprinted" cavities designed to match the shape and chemical properties of a target analyte. This design creates highly specific binding, analogous to antibody–antigen recognition.

MIPs are typically optimized for particular solvent systems that preserve the integrity of the imprinted binding sites. They are especially valuable in forensic toxicology for detecting targeted compounds such as opioids or designer drugs. For example, a MIP cartridge engineered for fentanyl derivatives can selectively retain and isolate these drugs even from highly complex biological mixtures. The high specificity of MIPs minimizes matrix interference, producing cleaner extracts and enhancing the sensitivity of downstream analyses. An overview of cartridge types employed in forensic toxicology, including their main features and applications, is presented in Table 6.13.

6.3.6 Workflow of SPE

The SPE workflow is built upon a sequence of carefully designed steps, each of which ensures efficient analyte retention, purification, and recovery from complex biological matrices. A streamlined process is essential to improve reproducibility and minimize matrix interference in forensic and toxicological applications. The sequential steps involved in SPE are presented in Fig. 6.8.

Conditioning: The first step involves activating the sorbent bed (e.g., C18, SCX) with a strong organic solvent such as methanol, followed by equilibration with water or buffer. This prepares the surface for effective analyte interaction. For example, conditioning may involve 2 mL of methanol followed by 2 mL of phosphate buffer at pH 6.0.

Sample Loading: The biological specimen (e.g., urine, blood) is introduced onto the cartridge. Adjusting the pH to match analyte chemistry is critical for instance, loading 1 mL of acidified urine at pH 2 onto a C18 cartridge to optimize drug retention.

Table 6.13 Comparison of SPE cartridge types commonly used in forensic toxicology

Cartridge type	Packing material	Specification	Solvent compatibility		Forensic applications
			Conditioning/wash	Elution	
Reversed-Phase (RP) Cartridges	C18 (Octadecyl silica)	18-carbon hydrophobic chains for strong non-polar interactions	Methanol → Water or buffer (pH-adjusted)	Organic solvents (methanol, acetonitrile, ethyl acetate)	Opioids (fentanyl, morphine), cannabinoids (THC-COOH), benzodiazepines
	C8 (Octyl silica)	Moderate hydrophobicity, ideal for less non-polar analytes			Cocaine metabolites, amphetamines (weaker retention than C18)
	C2/C4	Shorter chains for polar-neutral compounds			Protein and Peptides analysis
Ion-Exchange (IE) Cartridges	SCX (Strong Cation Exchange)	Sulfonic acid ($–SO_3^-$) for basic drugs (pH < pKa)	Methanol → pH-adjusted buffer (e.g., phosphate)	High-ionic-strength buffers (NH_4OH for SCX)	Amphetamines, phencyclidine (PCP), opioids (at low pH)
	SAX (Strong Anion Exchange)	Quaternary amine ($–NR_3^+$) for acidic drugs (pH > pKa)			Barbiturates, NSAIDs (e.g., ibuprofen)
	WCX (Weak Cation Exchange)	Carboxylic acid (–COOH) for selective basic drug retention			Protein and Peptides analysis

(continued)

Table 6.13 (continued)

Cartridge type	Packing material	Specification	Solvent compatibility		Forensic applications
			Conditioning/wash	Elution	
Mixed-Mode Cartridges	C18/SCX	Combines RP and cation exchange (e.g., Oasis MCX)	Methanol → Water → Buffer	Organic solvent + acid/base (e.g., 2% NH_4OH in methanol)	Broad spectrum (opioids, cocaine, benzodiazepines)
	C18/SAX	RP + anion exchange (e.g., Oasis MAX)			Acidic drugs (e.g., benzoylecgonine)
Normal-Phase (NP) Cartridges	Silica (Si–OH)	Polar for hydrophilic interactions	Non-polar solvents (hexane, chloroform)	Polar solvents (methanol, acetone)	LSD, psilocybin (polar hallucinogens)
	Florisil ($MgSiO_3$)	For highly polar compounds			Synthetic cannabinoids, pesticides
Polymer-Based Cartridges	HLB (Hydrophilic–Lipophilic Balance)	Copolymer (e.g., divinylbenzene) for broad-spectrum retention	Methanol → Water	Organic solvents (acetonitrile, methanol)	New psychoactive substances (NPS), poly-drug mixtures
	DVB (Divinylbenzene)	For non-polar analytes (PAHs, pesticides)			

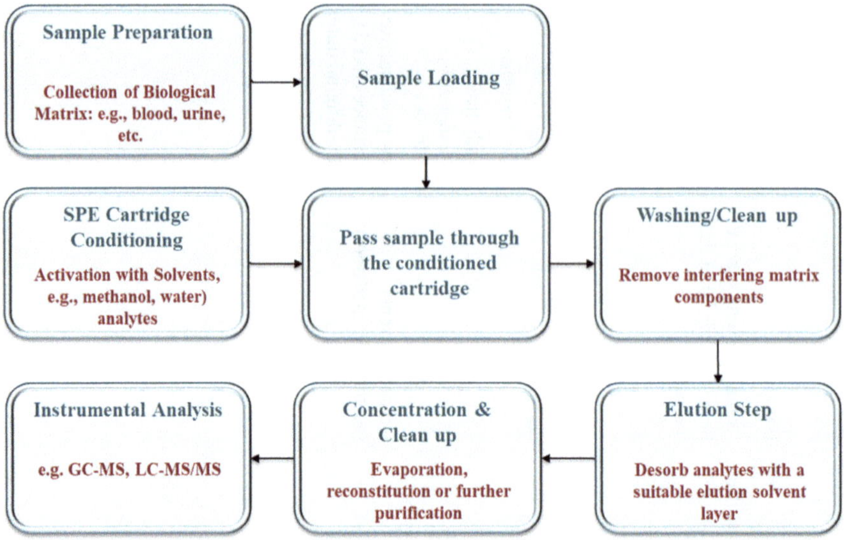

Fig. 6.8 Flow diagram illustrating the key steps in the SPE process

Washing: Unwanted matrix components such as salts and proteins are selectively removed with mild aqueous-organic solvents. For example, a 2 mL wash with 0.1 M acetic acid can eliminate residual impurities while leaving analytes bound.

Drying (Optional): Residual moisture is sometimes removed, often under vacuum, to improve elution efficiency.

Elution: Target analytes are released using a stronger solvent system tailored to their chemical nature, such as methanol with ammonium hydroxide (98:2, v/v) for basic drugs.

Evaporation/Reconstitution: Eluates are concentrated under a nitrogen stream and reconstituted in a solvent compatible with downstream analysis.

Analysis: The final extract is subjected to advanced instrumentation such as LC–MS (for polar analytes), GC–MS (for volatiles), or immunoassays.

6.3.7 Advantages and Limitations of SPE

SPE is widely recognized as one of the most effective sample preparation techniques in forensic toxicology, offering a balance of selectivity, sensitivity, and efficiency. One of its greatest strengths is the ability to perform selective cleanup, where analytes

Table 6.14 Comparative overview of SPE and LLE

Parameter	SPE	LLE
Solvent use	5–10 mL	10–50 mL
Automation	High (96-well plates)	Low (manual phase separation)
Matrix clean up	Excellent	Moderate
Selectivity	High (sorbent-specific)	Low (solvent-dependent)

of interest can be isolated from complex biological matrices by fine-tuning the interaction between the sorbent and the analyte. This leads to improved accuracy and reliability in downstream analyses.

Another advantage of SPE lies in its reduced solvent consumption. Unlike LLE, which often requires large volumes of organic solvents, SPE relies on smaller amounts, making it more cost-efficient over time and environmentally safer. Furthermore, SPE is automation-friendly; many laboratories employ automated systems that standardize the process, minimize operator-related variability, and improve reproducibility, a critical factor in forensic and regulatory settings. By allowing pre-concentration of analytes, SPE also boosts the sensitivity of advanced analytical techniques such as LC–MS/MS and GC–MS. Its versatility is enhanced by the wide range of sorbents available, enabling applications across analytes with diverse chemical properties, from non-polar drugs to highly polar metabolites. Details highlighting the strengths and drawbacks of SPE versus LLE in forensic toxicology are presented in Table 6.14.

Despite these advantages, SPE is not without limitations. Cost is a notable concern, as cartridges and automated platforms are more expensive compared to traditional methods like LLE. Method development can also be challenging, requiring careful selection of sorbent materials, pH conditions, and solvent systems to achieve optimal recovery. Additionally, capacity limitations exist; overloading the sorbent bed may lead to analyte breakthrough and reduced extraction efficiency. Finally, while SPE greatly reduces matrix interference, certain complex biological samples may still cause ion suppression or enhancement in mass spectrometry.

6.4 Solid-Phase Microextraction (SPME)

Solid-phase microextraction (SPME) is a versatile sample preparation technique that combines extraction and pre-concentration into a single, efficient process [33]. Designed to streamline rapid sample handling both in the laboratory and in the field, SPME has become increasingly popular over the last two decades for use with gas chromatography (GC) and liquid chromatography (LC) analyses [34]. Its simple design and operational ease have made it suitable for automation and high-throughput applications in both chromatographic platforms.

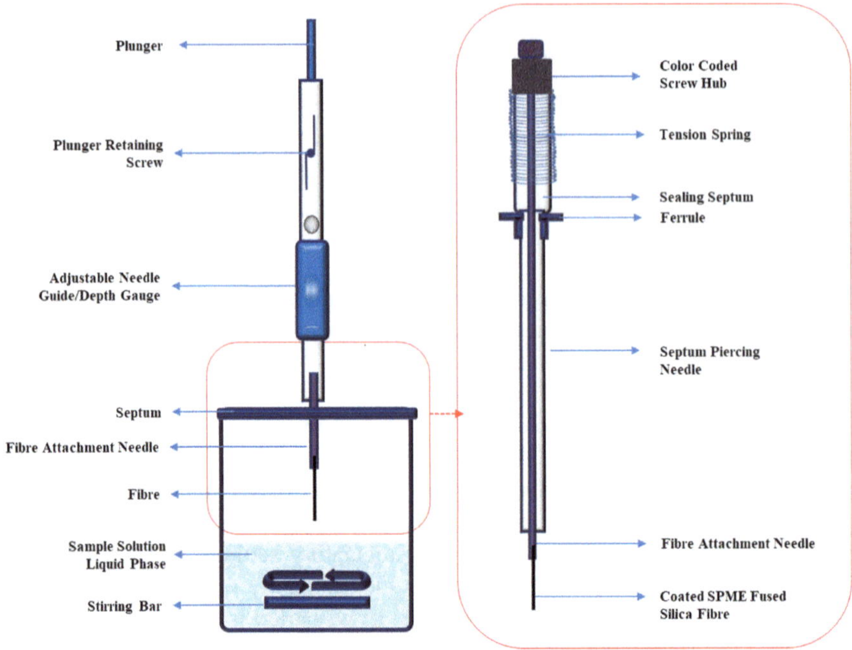

Fig. 6.9 Schematic overview of the SPME process

Developed in the early 1990s by Professor Janusz Pawliszyn at the University of Waterloo, Canada, SPME integrates sampling, extraction, concentration, and sample introduction into one streamlined workflow. The technique is solvent-free, employing a fiber constructed from fused silica or stainless steel coated with a thin polymer layer that acts as the extracting medium. This fiber is mounted on a syringe-like device, which facilitates direct analyte collection from various matrices and their subsequent introduction into chromatographic instruments. The method relies on partitioning equilibrium, where analytes distribute between the sample matrix (liquid, gas, or solid) and the fiber coating [35]. Initially promoted as an environmentally friendly alternative to conventional extraction techniques, SPME has become widely adopted in forensic toxicology due to its sensitivity, simplicity, and minimal solvent usage. Figure 6.9 depicts the schematic representation of the SPME process.

6.4.1 Theory and Principles of SPME

SPME is based on the principle of analyte partitioning between a sample and a stationary phase coated on a fiber. When the fiber is exposed to a sample either liquid or vapor analytes migrate onto the fiber coating according to their affinity until equilibrium is achieved [35, 36].

Fiber Coatings: The fiber's coating can be either a liquid phase immobilized on a solid support or a porous solid material. Common examples include polydimethylsiloxane (PDMS) for non-polar analytes and polyacrylate (PA) for polar compounds.

Equilibrium Partitioning: The amount of analyte absorbed by the fiber is determined by the fiber-sample partition coefficient (K_fs), which governs the distribution of analytes between the sample matrix and the fiber coating.

Modes of Operation: SPME can be carried out in two main modes

- *Direct Immersion (DI-SPME)*: The fiber is directly immersed into the liquid sample.
- *Headspace (HS-SPME)*: The fiber is exposed to the vapor phase above the sample, ideal for volatile or semi-volatile compounds.

Mechanisms of Extraction

- *Absorption (Partitioning)*: Analytes dissolve into a liquid polymer coating (e.g., PDMS), following Henry's Law (gas phase) or octanol–water partition coefficients (Log P, liquid phase).
- *Adsorption (Surface Binding)*: Analytes bind to solid sorbents (e.g., carbon, divinylbenzene), typically described by the Langmuir isotherm.

Key Equation

$$C_f = \frac{K_{fs} \cdot C_s \cdot V_s}{V_f + K_{fs} \cdot V_s}$$

where

C_f = analyte concentration on fiber.

C_s = analyte concentration in sample.

K_{fs} = fiber-sample partition coefficient.

V_f = fiber volume.

V_s = sample volume.

The types of commercially available fiber coatings and their typical applications are summarized in Table 6.15. The key components of a typical SPME setup are summarized in Table 6.16.

Table 6.15 Commonly used fiber coatings in SPME

Coating	Polarity	Target analytes	Max Temp (°C)
PDMS	Non-polar	Hydrocarbons, PAHs	300
PA	Polar	Phenols, amines	320
CAR/PDMS	Bipolar	Volatiles (e.g., pesticides)	280
DVB/CAR/PDMS	Mixed-mode	Broad-spectrum drugs	270

Table 6.16 Essential components of a typical SPME setup

Component	Description
Fiber core	Fused silica or metal wire (mechanically robust)
Coating material	· Non-polar: PDMS (polydimethylsiloxane)\ · Polar: PA (polyacrylate) · Mixed: PDMS/DVB (divinylbenzene)
Protective needle	Stainless steel sheath for fiber retraction

6.4.2 Extraction Kinetics

- The rate and efficiency of analyte extraction in SPME depend on several key factors that influence how quickly equilibrium is achieved and how effectively target compounds are captured.
- **Diffusion Rates**: The movement of analytes through the sample matrix to the fiber coating.
- **Surface Area and Coating Thickness**: A larger surface area and thinner coating can promote faster equilibration, though there is often a trade-off with capacity.
- **Temperature**: Increased temperature usually enhances analyte diffusion and can shift partitioning equilibria, though care must be taken to avoid degradation of thermo-labile compounds.
- **Stirring and Agitation**: Agitation can significantly reduce the time needed to reach equilibrium by minimizing the boundary layer around the Fiber.

Once the analytes have been extracted, the fiber is transferred to an analytical instrument:

- **Thermal Desorption**: Commonly used with GC, where the Fiber is inserted into a hot injector, releasing the analytes by heat.
- **Solvent Desorption**: For LC applications, the Fiber is placed in a small volume of solvent to release the analytes into the mobile phase.

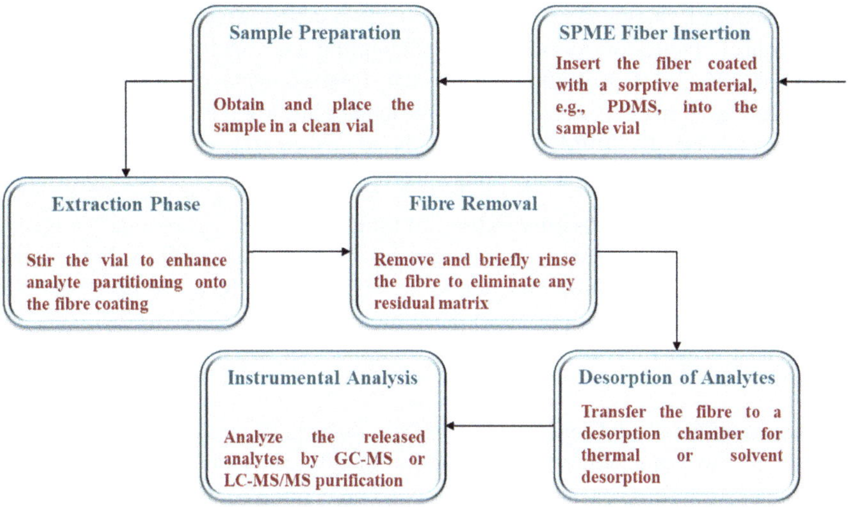

Fig. 6.10 Flow chart that outlines the key steps involved in SPME

6.4.3 Workflow of SPME

Figure 6.10 presents a flow chart outlining the key steps involved in SPME-based urine analysis.

- **Sample Preparation**: A portion of urine is transferred into a sealed vial. The pH is adjusted, typically to around 10, to favor the non-ionized form of amphetamines and enhance extraction efficiency.
- **Fiber Exposure**: A SPME fiber, coated with an appropriate phase such as PDMS or a mixed-mode coating, is positioned in the headspace above the urine sample.
- **Extraction**: Analytes migrate from the urine or its headspace into the fiber coating. Gentle agitation is applied to improve mass transfer and accelerate equilibrium between the sample and fiber.
- **Desorption**: The fiber is subsequently inserted into the injector of a gas chromatograph. Thermal desorption releases the analytes, which are then analyzed using GC–MS for identification and quantification.

This sequence ensures efficient, reproducible extraction while minimizing sample handling and solvent use.

6.4.4 Forensic Toxicology Applications

SPME has become an invaluable tool in forensic toxicology, providing a rapid, sensitive, and solvent-minimized approach for detecting and quantifying a wide array of substances. Early applications of in vivo SPME enabled researchers to investigate the pharmacokinetics of drugs by directly sampling analytes from the bloodstream of animals [36, 37]. For example, fibers coated with polypyrrole and polyethylene glycol were used to extract diazepam and its metabolites from beagle dogs, followed by analysis using LC–MS/MS. To accurately relate the extracted amounts to in vivo concentrations, both external and kinetic calibration approaches were employed.

Analysis of Volatile and Semi-volatile Compounds: SPME is particularly effective in detecting volatile substances such as ethanol, solvents, and certain volatile drugs. The headspace mode allows analytes to be isolated from blood or urine samples with minimal sample handling. In forensic fire investigations, SPME facilitates the detection of accelerants by sampling the headspace of fire debris, aiding in arson analysis.

Extraction of Non-volatile Compounds: For non-volatile analytes, including pharmaceuticals and metabolites, SPME offers an efficient method to extract these compounds from complex biological matrices. Direct immersion SPME is commonly used to detect drugs such as benzodiazepines or opiates in blood, urine, or tissue samples. Additionally, SPME can capture biomarkers from tissues, hair, or nails, providing crucial information on exposure to toxins.

Multi-residue and Screening Applications: The high sensitivity and broad applicability of SPME make it suitable for multi-residue screening. A single fiber can often capture a wide range of analytes simultaneously, including drugs, metabolites, and environmental contaminants. Its compatibility with automation enables forensic laboratories to process multiple samples efficiently, enhancing throughput while minimizing operator-dependent variability.

6.4.5 Advantages of SPME

- **Minimal Solvent Use**: Environmentally friendly and cost-effective.
- **Integrated Sampling**: Combines extraction and pre-concentration in one step, reducing contamination risk.
- **Versatility**: Suitable for both volatile and non-volatile analytes across diverse matrices.
- **Automation-Friendly**: Easily adapted to high-throughput systems, improving reproducibility and efficiency.

The key advantages of SPME compared to conventional extraction techniques are summarized in Table 6.17.

Table 6.17 Advantages of SPME over traditional methods

Parameter	SPME	LLE/SPE
Solvent use	None	Large volumes (toxic waste)
Speed	5–30 min (automated)	1–2 h
Sensitivity	ppt–ppb level	Ppb-ppm
Portability	Field-deployable (e.g., crime scenes)	Lab only

6.4.6 Limitation of SPME

SPME is a groundbreaking, solvent-free technique that streamlines sample preparation for forensic and analytical applications, yet it has some inherent limitations.

- **Limited Capacity**: The Fiber has a finite coating volume, which can limit the extraction of very low-concentration analytes.
- **Matrix Effects**: Complex biological samples may require additional optimization to minimize interference.
- **Method Development**: Requires careful selection of Fiber type, extraction mode, and optimization of extraction time and temperature for each analyte.

Despite these limitations, SPME remains highly valuable. Its equilibrium-based extraction preserves labile or short-lived metabolites, supports both broad-spectrum and targeted analyses, and pairs effectively with advanced platforms like GC–MS. In forensic toxicology, it excels at detecting volatile substances (e.g., ethanol, inhalants) and trace-level non-volatile drugs (e.g., opioids, benzodiazepines), making it an essential tool in modern analytical workflows.

6.5 Dispersive Liquid–Liquid Microextraction (DLLME)

Dispersive liquid–liquid microextraction (DLLME) is a miniaturized, highly efficient sample preparation technique developed to isolate analytes from aqueous samples using minimal amounts of organic solvents [38, 39]. Its design emphasizes speed, simplicity, and environmental safety, making it particularly suitable for trace-level analysis in complex matrices such as biological fluids and environmental samples. Over recent years, DLLME has gained prominence in forensic toxicology due to its rapid extraction kinetics, high enrichment factors, and significantly reduced solvent consumption compared to traditional LLE methods [40].

The technique is valued not only for its efficiency but also for its operational practicality. Its use of small solvent volumes enhances laboratory safety and minimizes environmental impact, while the straightforward procedure allows for rapid sample processing without compromising analytical performance [41]. Despite being solvent-based, DLLME's effectiveness largely depends on the partitioning

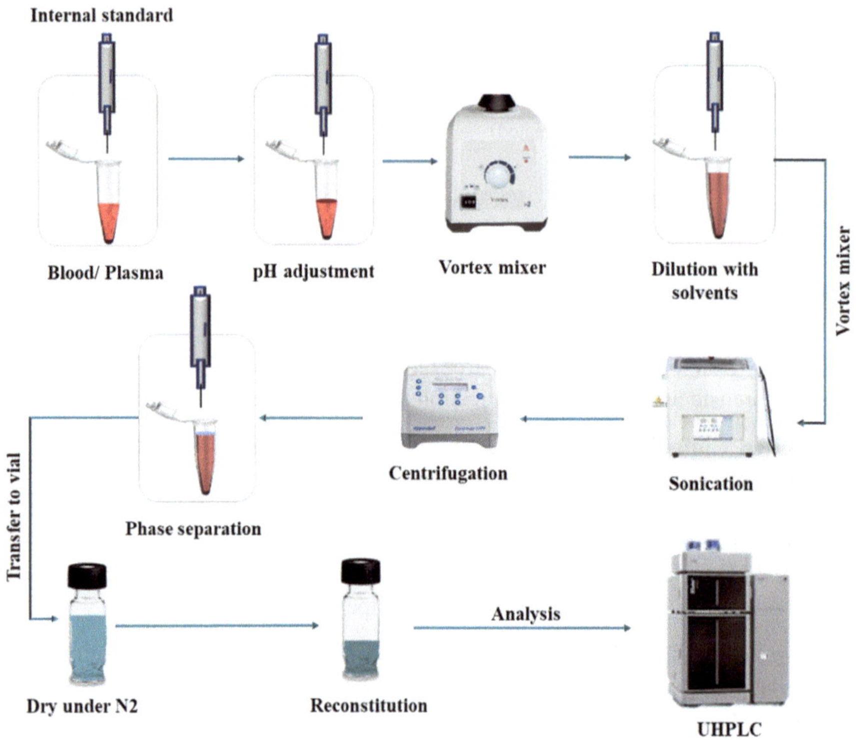

Fig. 6.11 Schematic illustration of the DLLME process

behavior of analytes between the aqueous sample and the extraction solvent, typically expressed by the partition coefficient (K_D). Proper selection of solvents and understanding analyte properties are critical to achieving optimal recovery. Figure 6.11 presents a schematic overview of the DLLME procedure for aqueous sample extraction.

6.5.1 Basic Principle

The core principle of DLLME is the rapid formation of a fine, cloudy solution when a mixture of an extraction solvent (immiscible with water) and a disperser solvent (miscible with both water and the extraction solvent) is injected into an aqueous sample. This dispersion creates a large interfacial area between the aqueous phase and extraction solvent, dramatically enhancing the mass transfer of analytes into the solvent droplets. The critical parameters affecting DLLME performance are summarized in Table 6.18.

Table 6.18 Key parameters influencing the efficiency of DLLME

Factor	Role
Extraction Solvent	High density, low solubility in water (e.g., chloroform, carbon tetrachloride)
Disperser solvent	Miscible with both phases (e.g., acetone, methanol) to stabilize droplets
pH adjustment	Optimizes ionization state of analytes (e.g., basic drugs extracted at low pH)
Salt addition	Enhances extraction efficiency via salting-out effect

Key Mechanisms

- Forma*tion of Microdroplets*: The disperser solvent, commonly acetone, promotes uniform dispersion of the extraction solvent often a chlorinated solvent or an ionic liquid within the aqueous sample. This produces tiny droplets, which maximize contact between the phases and facilitate rapid analyte transfer.
- *Partition Coefficient*: The distribution of analytes between the aqueous phase and extraction solvent is governed by their hydrophobicity and solubility, expressed as the partition coefficient (K_D). Analytes with higher affinity for the extraction solvent are efficiently concentrated within the microdroplets, allowing subsequent analysis.

6.5.2 Applications in Forensic Toxicology

DLLME has become an increasingly valuable tool in forensic toxicology, offering rapid and efficient isolation of drugs, toxicants, and other analytes from complex biological matrices. Its high enrichment capability and minimal solvent use make it particularly suitable for challenging samples such as blood, urine, saliva, and various tissue types. By generating fine microdroplets of the extraction solvent dispersed within the sample, DLLME ensures rapid analyte transfer and concentration, improving the sensitivity of subsequent analytical techniques such as gas or liquid chromatography coupled with mass spectrometry.

This versatility allows DLLME to be applied in diverse forensic scenarios, including the detection of therapeutic drugs, illicit substances, and environmental toxins. Its simplicity, speed, and reproducibility have made it a practical choice for both routine forensic analysis and research-oriented investigations. Table 6.19 highlights how DLLME is applied for the extraction and analysis of various toxicants and drugs in forensic matrices.

6.5.3 Workflow of DLLME

The workflow of DLLME is simple, rapid, and highly efficient, as illustrated in Fig. 6.12.

Table 6.19 Key applications of DLLME in forensic toxicology

Drug class	Examples	Matrix	Extraction solvents	Detection method
Benzodiazepines	Diazepam, Lorazepam	Blood, Urine	Chloroform, Acetonitrile	LC–MS/MS
Opiates	Morphine, Codeine	Urine, Saliva	Chlorobenzene, Methanol	GC–MS
Antidepressants	Amitriptyline, Fluoxetine	Plasma	Carbon tetrachloride	LC–MS
Stimulants	Cocaine, Methamphetamine	Blood	Tetrachloroethane	GC–MS
NSAIDs	Ibuprofen, Diclofenac	Urine, Serum	Chloroform	HPLC–UV
Organophosphates	Malathion, Parathion	Blood	Dichloromethane	GC-ECD

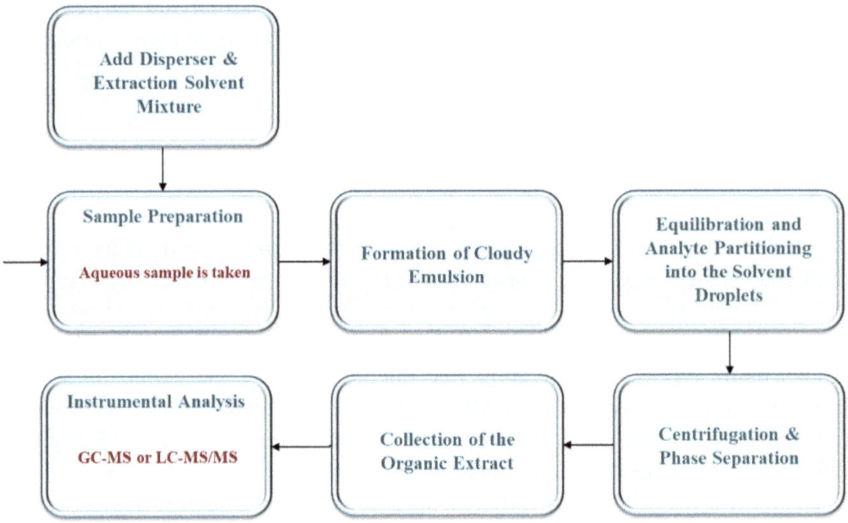

Fig. 6.12 Flowchart illustrating the key steps in DLLME

- **Selection of Solvents**: The process begins with choosing an appropriate extraction solvent, typically a chlorinated compound such as chlorobenzene, carbon tetrachloride, or chloroform, and a disperser solvent that is miscible with both water and the extraction solvent, such as acetone, methanol, or acetonitrile.
- **Injection and Dispersion**: A mixture of the extraction and disperser solvents is swiftly injected into the aqueous sample containing the target analytes. This action forms a fine, cloudy emulsion, enhancing the contact surface area between the extraction solvent and the sample.

- **Centrifugation and Phase Separation**: After a brief equilibrium period, the sample is centrifuged to separate the extraction solvent droplets, which now contain the concentrated analytes.
- **Analysis**: The sedimented extraction phase is then collected and subjected to analytical techniques such as GC–MS or LC–MS/MS for detection and quantification.

6.5.4 Advantages of DLLME

DLLME offers numerous benefits that make it an attractive choice for forensic and analytical applications.

- **High Extraction Efficiency and Enrichment Factor**: The rapid formation of a cloudy emulsion generates extremely fine droplets of the extraction solvent, dramatically increasing the contact surface area with the aqueous sample. This enhanced interface promotes efficient mass transfer of analytes, allowing for the concentration of target compounds even at trace levels.
- **Rapid Extraction Process**: DLLME achieves phase equilibrium within a few minutes, enabling timely analysis, which is particularly critical in forensic investigations.
- **Minimal Solvent**: Minimal solvent consumption further distinguishes DLLME from traditional LLE methods. Only small volumes of solvent are required, aligning with green chemistry principles, reducing operational costs, lowering environmental impact, and improving laboratory safety.
- **Simplicity and Ease of Implementation**: The method's simplicity and ease of implementation make it accessible for routine use. DLLME does not require sophisticated or expensive equipment, and its straightforward workflow can be quickly mastered by laboratory personnel.
- **Versatility Across Diverse Matrices**: Versatility across diverse matrices enhances its appeal. DLLME can efficiently extract analytes from biological fluids, environmental samples, and food products, supporting a wide range of forensic applications.

6.5.5 Disadvantages of DLLME

Despite its advantages, DLLME has some limitations.

- Its reliance on partition coefficients means that analytes must preferentially distribute into the extraction solvent; otherwise, recoveries may be low.
- Emulsion stability issues can occasionally arise, complicating the separation of the solvent from the aqueous phase and potentially affecting reproducibility.

- DLLME is less effective for highly hydrophilic or extremely volatile compounds, as these analytes may not partition efficiently into the organic solvent or could be lost during the extraction process.
- Additionally, method optimization is crucial: parameters such as solvent type and ratio, sample pH, temperature, and centrifugation conditions must be carefully tailored to each analyte and matrix, which can be time-consuming.
- Matrix interferences in complex samples can also co-extract with target analytes, sometimes necessitating additional cleanup or more selective detection techniques.

Overall, DLLME is a fast, efficient, and environmentally friendly microextraction technique that has found widespread utility in forensic toxicology. Its high enrichment capabilities, low solvent use, and adaptability to diverse matrices make it a powerful tool. However, its performance is closely tied to analyte properties and sample characteristics, and careful optimization is essential. Understanding both its strengths and limitations is critical for effectively applying DLLME in routine and specialized forensic analyses.

6.6 Microextraction by Packed Sorbent (MEPS)

MEPS is a miniaturized form of SPE designed to simplify and accelerate sample preparation while significantly reducing solvent usage [42]. Based on the principles of selective adsorption and partitioning, MEPS employs a small quantity of sorbent material typically 1–3 mg housed within a syringe barrel or pipette tip [43]. Analytes of interest bind to the sorbent, after which they are efficiently eluted in a concentrated form for subsequent analysis.

Owing to its speed, efficiency, and compatibility with automation, MEPS has gained increasing attention in forensic toxicology. The method is particularly suited for scenarios where only limited sample volumes are available, such as blood, urine, saliva, or postmortem specimens, and where analytes (e.g., drugs, metabolites, or toxic substances) often occur at trace concentrations [44–46].

6.6.1 Fundamental Concept

MEPS shares its conceptual basis with conventional SPE conditioning, adsorption, washing, and elution but operates on a much smaller scale [47–49].

- **Conditioning**: The sorbent is first activated using a small amount of solvent, often followed by water or buffer, to ensure effective analyte binding.

- **Sample Loading**: A small sample volume (100 μL–a few mL) is aspirated into the syringe, where analytes partition between the matrix and sorbent. Retention depends on hydrophobic, hydrogen bonding, or ion-exchange interactions, depending on the chemistry of the sorbent (e.g., C18).
- **Washing**: Interfering matrix components are removed using a carefully selected wash solvent, ensuring that analyte–sorbent interactions are maintained.
- **Elution**: A minimal volume of solvent is used to release analytes, yielding a clean, concentrated extract suitable for instrumental analysis, most commonly LC–MS/MS.

The schematic representation of MEPS is shown in Fig. 6.13. The design of MEPS makes it highly adaptable to automated robotic systems. By programming repetitive aspiration and ejection cycles, laboratories can achieve high-throughput and reproducible extractions, a critical requirement in forensic workflows.

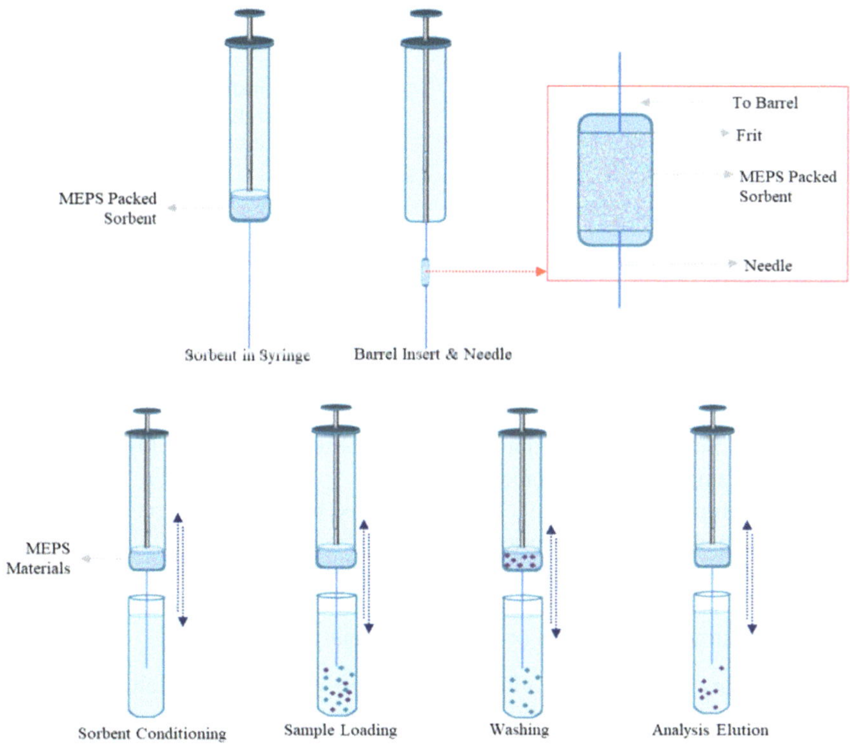

Fig. 6.13 Microextraction by Packed Sorbent (MEPS): Schematic representation

Table 6.20 Applications of MEPS in forensic toxicology

Drug class/analyte	Typical sorbent used	Sample type	Solvent(s) for conditioning/elution	Application example
Basic Drugs (Amphetamines, Opioids)	Mixed-mode (C18 + cation-exchange)	Urine, Blood	Condition: Methanol + Water; Elute: Methanol with 0.1% formic acid	Detection of amphetamines in urine for drug abuse testing
Acidic Drugs (NSAIDs, Barbiturates)	C18 or anion-exchange	Blood, Plasma	Condition: Methanol, Water; Elute: Acetonitrile with acid	Quantification of NSAIDs in plasma for toxicological analysis
Neutral Compounds (Benzodiazepines, Steroids)	C18 reversed-phase	Serum, Urine	Condition: Methanol + Water; Elute: 100% Methanol or Acetonitrile	Extraction of benzodiazepines from serum in forensic case work
Multi-residue/ metabolites	Mixed-mode, HLB	Urine	Condition: Organic modifier and buffer; Elute: Optimized solvent blend	Comprehensive metabolite profiling in forensic toxicology

6.6.2 *Operational Considerations in Forensic Applications*

MEPS offers several practical benefits:

- **Automation and Throughput**: Well-suited for robotic platforms, ensuring speed, reproducibility, and reduced operator variability.
- **Minimal Solvent Use**: Requires only microliters of solvent, lowering costs, environmental impact, and laboratory risks.
- **Small Sample Volumes**: Effective with limited biological material, particularly useful in pediatric or postmortem cases.
- **Matrix Clean-Up and Enrichment**: The sorbent's high surface-to-volume ratio promotes selective analyte capture and removal of interferences, producing cleaner extracts for GC–MS or LC–MS/MS analysis.

The primary forensic applications of MEPS are summarized in Table 6.20.

6.6.3 *Comparison of MEPS Versus Traditional SPE*

MEPS and traditional SPE share the same fundamental principles but differ significantly in scale, efficiency, and practicality. MEPS requires only a few milligrams of

Table 6.21 Comparative features of MEPS and conventional SPE

Parameter	MEPS	Traditional SPE
Solvent use	20–50 μL	5–50 mL
Sample volume	10–1000 μL	1–10 mL
Reusability	100 + times	Single-use
Automation	Fully compatible	Limited
Cost per sample	Low	High
Matrix tolerance	Moderate (needs pre-treatment)	High (handles dirty samples)

sorbent and microliter volumes of solvent, making it cost-effective, environmentally friendly, and well-suited for small or limited biological samples. Its compatibility with automation further enhances reproducibility and throughput in forensic laboratories. In contrast, traditional SPE uses larger sorbent cartridges and higher solvent volumes, which increase costs and waste. While SPE remains highly versatile, MEPS offers a faster, cleaner, and greener alternative for modern forensic applications. Table 6.21 compares the key attributes of MEPS with traditional SPE approaches.

6.6.4 Workflow of MEPS

The workflow of MEPS follows a streamlined sequence of steps that integrates conditioning, extraction, washing, and elution into a miniaturized format. This process ensures efficient analyte recovery from complex biological matrices while requiring only minimal sample and solvent volumes. Figure 6.14 provides a schematic flow diagram outlining the workflow of MEPS from sample loading to instrumental analysis.

Key Steps

- **Biological Sample**: The process begins with the introduction of a biological sample, such as blood, urine, or saliva.
- **Conditioning of MEPS Cartridge**: The cartridge is preconditioned with suitable solvents (e.g., methanol followed by water or buffer) to activate the sorbent.
- **Sample Loading**: The sample is aspirated through the cartridge, allowing analytes to interact with the sorbent via hydrophobic, ionic, or hydrogen-bonding mechanisms.
- **Washing Step**: Interfering matrix substances are removed using a wash solvent without dislodging the analytes.
- **Elution Step**: A minimal volume of solvent releases the analytes in a concentrated form.
- **Collection and Instrumental Analysis**: The eluted extract is collected and introduced into LC–MS/MS or GC–MS for separation and quantification.

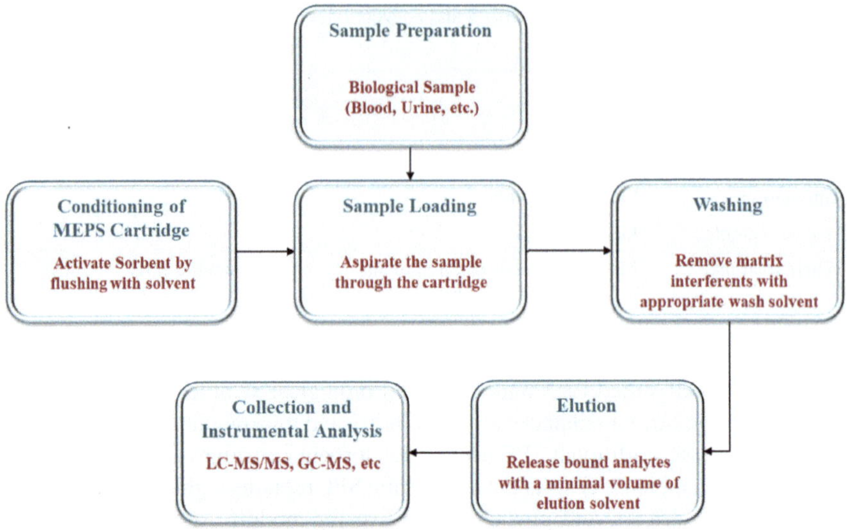

Fig. 6.14 Flow diagram illustrating the key steps in the MEPS workflow

6.6.5 Advantages of MEPS

MEPS has gained considerable attention in forensic toxicology and other analytical fields due to its ability to simplify and accelerate sample preparation while maintaining high efficiency. Its design as a miniaturized version of traditional SPE brings several significant advantages that make it particularly valuable for laboratories handling complex biological matrices and working with limited sample volumes.

Miniaturization and Efficiency: One of the most prominent strengths of MEPS lies in its miniaturization. Unlike conventional SPE, which requires large cartridges packed with hundreds of milligrams of sorbent, MEPS uses only a few milligrams. This reduction allows it to operate effectively with very small sample volumes, making it highly suitable for forensic cases where only a limited quantity of blood, urine, or other biological material is available. The smaller dimensions of the packed bed also promote rapid extraction kinetics, enabling faster binding and release of analytes without compromising recovery. In busy laboratories, this efficiency translates into quicker turnaround times and improved productivity.

Enhanced Analyte Enrichment: Another key benefit of MEPS is its ability to enrich analytes present at trace concentrations. Forensic toxicology frequently involves detecting drugs or toxicants in minute quantities, often at the edge of instrument detection limits. The selective retention of analytes on the sorbent, followed by elution into a concentrated solvent phase, allows MEPS to achieve significant enrichment factors. This capability is particularly critical when analyzing postmortem samples or cases involving low-dose drug exposure, where the analyte concentration may be extremely low yet of great forensic importance.

Ease of Automation: MEPS is inherently designed for automation, offering seamless integration with robotic liquid-handling systems and autosamplers. Automation reduces manual intervention, thereby minimizing operator variability and the likelihood of error. This feature ensures greater reproducibility of results while enhancing throughput in high-demand forensic laboratories. With increasing reliance on automated platforms for sample preparation, MEPS provides an excellent fit for modern analytical workflows.

Reduced Solvent Consumption: A major limitation of traditional SPE is its high consumption of organic solvents, which not only increases operational costs but also generates hazardous waste. MEPS, by contrast, requires only minimal volumes of solvents during conditioning, washing, and elution steps. This reduction aligns well with green chemistry principles by lowering chemical waste and minimizing laboratory exposure to harmful solvents. From both environmental and financial perspectives, reduced solvent use makes MEPS a cost-effective and sustainable alternative.

Versatility Across Matrices: The versatility of MEPS further enhances its applicability in forensic toxicology. It can be employed for a wide range of biological samples, including blood, urine, saliva, and even tissue homogenates. This adaptability allows laboratories to streamline sample preparation across diverse case types without needing multiple extraction platforms. By effectively removing matrix interferences while concentrating analytes, MEPS proves to be a robust and flexible tool for various analytical challenges.

6.6.6 Limitations of MEPS

Despite its numerous advantages, MEPS also presents certain limitations that must be considered before adopting it as a routine sample preparation strategy.

Limited Sorbent Capacity: The miniaturization that makes MEPS efficient also limits the amount of sorbent available for analyte binding. When analyzing samples with high concentrations of interfering compounds, sorbent overloading can occur, leading to reduced analyte recovery or inconsistent results. Careful attention to sample complexity is therefore essential.

Optimization Requirement: MEPS performance depends heavily on method optimization, including the choice of sorbent type, solvent systems for conditioning and elution, and the number of extraction cycles. Developing and validating these parameters for each new application can be time-consuming. This requirement may pose a challenge in forensic laboratories where quick adaptability is often necessary.

Matrix Effects: Although MEPS effectively reduces matrix interference, complex biological fluids can still introduce challenges. Residual proteins, lipids, or salts may compete for binding sites or cause sorbent fouling, potentially lowering sensitivity and reproducibility. Addressing these issues may require additional cleanup steps or repeated extraction cycles.

Reusability and Maintenance: Some MEPS devices are designed to be reusable, which is advantageous from a cost perspective. However, repeated use may degrade the sorbent material or lead to fouling, thereby reducing extraction efficiency. Ensuring reliable performance across multiple uses requires careful monitoring and, in some cases, frequent replacement of cartridges.

Instrumentation Integration: While MEPS is highly compatible with modern LC–MS/MS and GC–MS platforms, integrating it into existing laboratory workflows may require additional adjustments or method modifications. This adaptation process can create initial barriers for laboratories already accustomed to traditional SPE protocols.

MEPS provides a powerful and efficient approach for forensic toxicology, offering advantages such as miniaturization, high analyte enrichment, compatibility with automation, reduced solvent use, and applicability to diverse biological matrices. These strengths make it an attractive alternative to conventional SPE for handling trace-level compounds in limited and complex samples. However, limitations including reduced sorbent capacity, the need for careful optimization, potential matrix effects, and challenges related to reusability and integration must be addressed to fully realize its potential. A balanced evaluation of these advantages and limitations enables forensic laboratories to determine whether MEPS is the most suitable choice for their analytical and operational needs.

6.7 Stir Bar Sorptive Extraction (SBSE)

Stir Bar Sorptive Extraction (SBSE) is an advanced sample preparation technique that combines the processes of extraction and pre-concentration into a single, efficient step. It makes use of a magnetic stir bar coated with a sorptive material, most commonly PDMS, which selectively traps analytes from complex matrices [50, 51]. This approach has gained recognition in forensic toxicology and environmental chemistry because of its ability to handle small sample volumes while providing high enrichment factors, essential for detecting compounds at trace levels.

6.7.1 Fundamental Principle

The underlying principle of SBSE relies on equilibrium partitioning. When a coated stir bar is immersed in a liquid sample and agitated, target analytes migrate from the bulk solution into the sorptive coating [52]. With time, a dynamic balance is established between the concentration of analytes in the sample and in the coating, allowing efficient extraction.

Sorbent Coating and Material: The coating, usually PDMS or a related polymer, provides a large surface area that attracts hydrophobic compounds. The choice of sorbent material plays a decisive role in determining the selectivity of extraction, as different coatings have distinct affinities for analytes based on chemical properties such as polarity and hydrophobicity.

Introduction into the Sample Matrix: The coated stir bar is placed into the biological or environmental sample commonly blood, urine, saliva, or water and stirred. Agitation is essential, as it reduces the stagnant boundary layer around the stir bar, ensuring efficient contact between the analytes and the sorptive surface.

Analyte Partitioning and Equilibrium: Extraction efficiency is governed by the partition coefficient (K) of each analyte between the liquid phase and the coating. A higher K value indicates stronger sorption into the coating. With continued stirring, analytes distribute until equilibrium is reached, meaning their concentration in the coating is directly proportional to their concentration in the sample.

Desorption and Analysis: Once extraction is complete, the stir bar is retrieved and analytes are released by either thermal desorption directly into the injector of a gas chromatograph or solvent desorption for liquid chromatography-based analysis. This dual compatibility with GC–MS and LC–MS/MS enhances the versatility of SBSE.

Factors Influencing Extraction:

- Temperature: Elevated temperatures improve diffusion rates but must be controlled to avoid analyte degradation.
- Stirring Rate: Faster agitation reduces boundary effects, speeding up equilibrium.
- Coating Properties: Thickness and polarity of the sorbent layer influence analyte capacity and selectivity.

The workflow of SBSE, including analyte partitioning and subsequent desorption, is illustrated in Fig. 6.15.

6.7.2 *Forensic Toxicological Applications of SBSE*

In forensic toxicology, SBSE has emerged as a valuable tool for isolating drugs, metabolites, toxins, and volatile compounds from complex biological matrices [53–56]. Its efficiency, combined with minimal solvent requirements, makes it well-suited for high-sensitivity casework where analytes are often present at very low levels.

Detection of Drugs of Abuse: SBSE is widely used to extract lipophilic drugs such as benzodiazepines, cannabinoids, and opioids. PDMS-coated stir bars have shown excellent performance in capturing these compounds from blood and urine. After

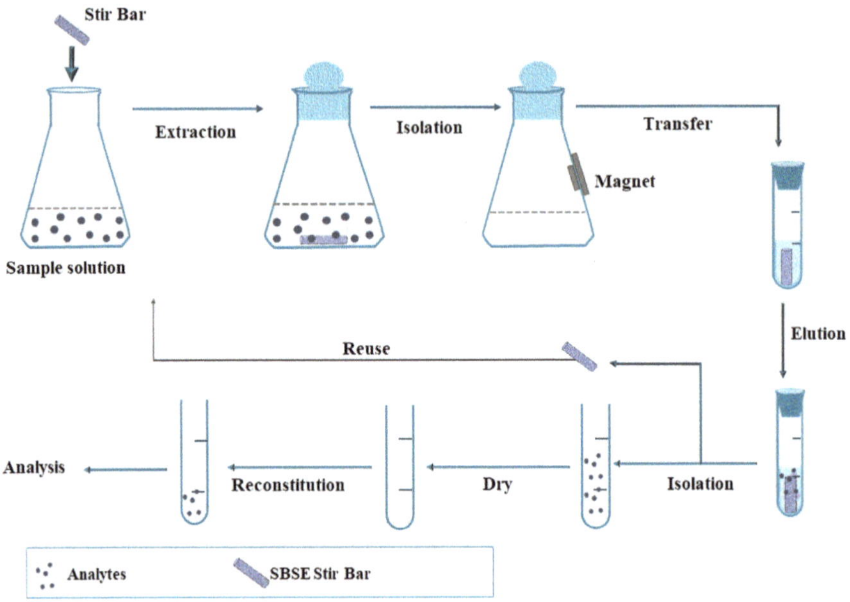

Fig. 6.15 Schematic representation of SBSE process

desorption, the concentrated analytes can be analyzed using GC–MS or LC–MS/MS, providing precise and reliable results.

Analysis of Toxins and Environmental Contaminants: In suspected poisoning cases, SBSE can isolate pesticides, industrial pollutants, and other organic toxicants from forensic samples. Its high enrichment capability ensures that even trace amounts of toxicants can be detected, supporting toxicological investigations where exposure levels are extremely low.

Volatile Organic Compounds (VOCs) in Postmortem Investigations: SBSE also plays a role in postmortem analysis, particularly in detecting VOCs released during decomposition or arson cases. The technique efficiently captures volatile and semi-volatile substances from biological fluids, providing investigators with crucial chemical evidence for cause-of-death assessments or crime scene reconstructions.

6.7.3 Workflow of SBSE

SBSE follows a systematic process that integrates extraction and pre-concentration of analytes from biological matrices. The technique is straightforward and effective, allowing sensitive detection of drugs, toxins, and other compounds commonly encountered in forensic toxicology. The sequential workflow of SBSE is illustrated in Fig. 6.16.

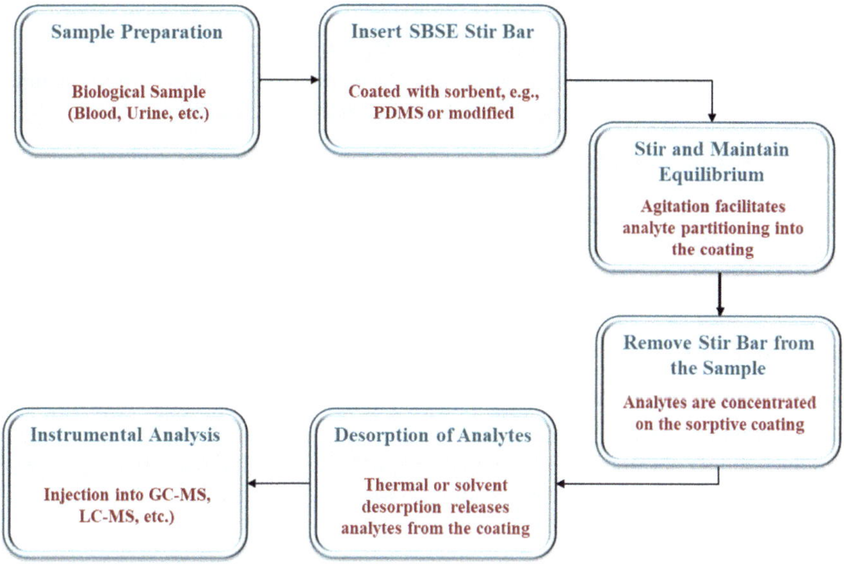

Fig. 6.16 Flow diagram outlining the key steps in SBSE

The overall workflow can be summarized in the following steps:

- **Biological Sample**: The procedure begins with a biological fluid such as blood or urine containing the analytes of interest.
- **Insertion of Stir Bar**: A stir bar coated with a sorptive material, usually PDMS, is placed into the sample.
- **Agitation and Equilibrium**: Stirring promotes efficient partitioning of analytes into the coating until equilibrium is achieved.
- **Stir Bar Removal**: After extraction, the stir bar is withdrawn, now enriched with concentrated analytes.
- **Desorption of Analytes**: The analytes are released via thermal desorption or solvent elution.
- **Instrumental Analysis**: Finally, the desorbed analytes are analyzed using GC–MS or LC–MS techniques.

6.7.4 Advantages of SBSE

SBSE stands out as a robust, eco-friendly, and highly sensitive technique for forensic toxicology, capable of isolating and enriching analytes from diverse matrices.

High Enrichment and Sensitivity

- The stir bar in SBSE is coated with a thick sorptive phase, providing a large surface area for analyte capture.

- This design allows significant pre-concentration of compounds, particularly those present at trace levels, resulting in lower detection limits.
- Such sensitivity is especially beneficial in forensic toxicology, where accurate detection of drugs or toxins at very low concentrations is critical.

Minimal Use of Solvents

- SBSE is a nearly solvent-free method, eliminating the need for large quantities of organic solvents.
- This reduces analysis costs, improves operator safety, and minimizes hazardous waste, aligning with sustainable "green chemistry" practices.

Integration of Extraction and Concentration

- Unlike conventional multi-step methods, SBSE combines extraction and concentration into one step.
- This not only shortens sample preparation time but also reduces opportunities for analyte loss or contamination, leading to more reliable results.

Ease of Automation

- The simplicity of the SBSE process makes it highly adaptable for automated systems.
- Automation improves reproducibility, minimizes manual labor, and enables high-throughput processing essential in forensic laboratories that routinely handle large sample volumes.

Versatility Across Different Matrices

- SBSE has been successfully applied to biological samples such as blood, urine, and hair, as well as environmental matrices.
- Its flexibility enhances its value as a universal tool for detecting drugs of abuse, environmental contaminants, and toxins in diverse forensic contexts.

6.7.5 Limitations of SBSE

Despite its many advantages, SBSE faces certain challenges that can affect its efficiency and reliability, particularly when applied to complex forensic samples.

Limited Extraction of Polar Analytes

- Standard coatings like PDMS are highly effective for non-polar compounds but show poor affinity for highly polar substances.
- Specialized coatings may be required, limiting the universal applicability of SBSE.

Extended Equilibration Times

- Achieving equilibrium between the analytes and sorbent can be slow, particularly for viscous or minimally agitated samples, reducing throughput.

Matrix Interferences

- Complex biological matrices may introduce interfering substances, leading to co-extraction of unwanted compounds and impacting analytical accuracy.

Potential for Coating Fouling

- Repeated use with complex samples can degrade or foul the coating, requiring frequent cleaning or replacement of stir bars, which adds cost and maintenance effort.

Challenges in Desorption Efficiency

- Both thermal and solvent-based desorption require careful optimization.
- Incomplete desorption can reduce analyte recovery, thereby affecting sensitivity and overall method performance.

SBSE is a sensitive, eco-friendly, and efficient technique for forensic toxicology, offering strong analyte enrichment with minimal solvent use and easy automation. However, limitations such as poor recovery of polar compounds, matrix interferences, and coating maintenance require careful consideration to ensure reliable application.

6.8 Supported Liquid Extraction (SLE)

Supported Liquid Extraction (SLE) is an advanced sample preparation technique that simplifies traditional LLE by incorporating a solid support material [57]. It is widely used in forensic toxicology for the efficient isolation of drugs, metabolites, and toxicants from complex biological matrices such as blood, urine, and tissue. By improving phase separation and minimizing matrix interferences, SLE enables accurate, sensitive, and reproducible analysis [58, 59].

6.8.1 Fundamental Concept

In SLE, the aqueous sample is first applied onto an inert solid support, typically diatomaceous earth or modified silica gel. The support retains the aqueous phase within its porous structure, distributing the analytes across a thin film. An immiscible organic solvent is then applied, extracting analytes based on their solubility and polarity. Key aspects include:

- The solid support holds the aqueous sample (blood, urine, or tissue extract) in a thin layer.
- Analytes partition into the organic solvent during contact, while interfering components such as proteins and salts remain on the support or in the aqueous

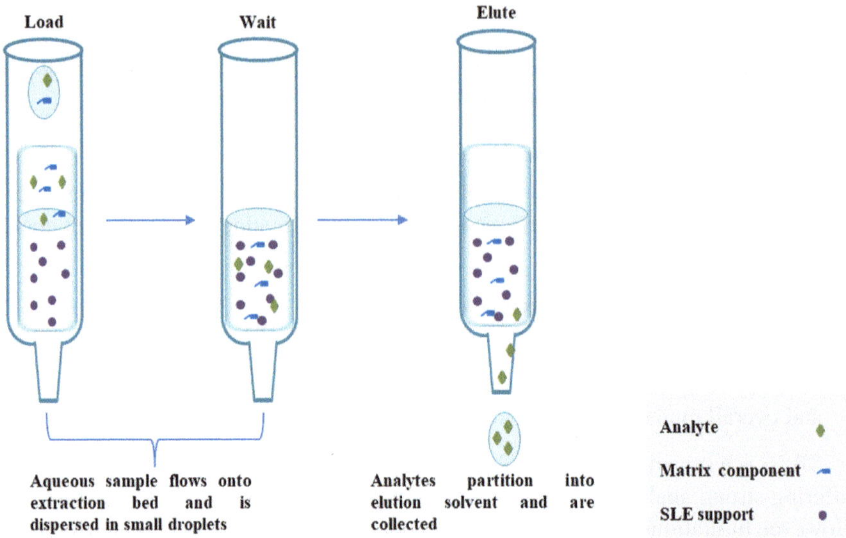

Fig. 6.17 Schematic representation of SLE

phase. This process produces a cleaner extract that is suitable for downstream analysis.

As depicted in Fig. 6.17, SLE facilitates efficient extraction and cleanup of analytes from complex biological matrices.

6.8.2 Mechanism of Action

The mechanism of SLE involves sequential steps that facilitate efficient analyte transfer from the aqueous sample to the organic solvent, ensuring high recovery and minimal matrix interference.

Sample Loading: The biological sample is loaded onto a cartridge containing the inert support, creating a uniform thin layer that maximizes surface area for interaction with the extraction solvent.

Solvent Contact and Partitioning: An immiscible organic solvent is introduced, percolating through the solid support and forming a transient interface with the aqueous layer. Analytes distribute into the organic phase according to their partition coefficients, with compounds more soluble in the organic phase preferentially migrating out of the aqueous matrix.

Phase Separation: The solid support reduces emulsion formation, a common limitation of conventional LLE, and facilitates rapid, efficient separation of the solvent containing the extracted analytes.

Concentration and Analysis: The enriched organic phase can be concentrated if necessary and then directly analyzed using advanced instrumentation such as liquid chromatography-mass spectrometry (LC–MS) or gas chromatography-mass spectrometry (GC–MS) for precise quantification and identification.

6.8.3 Workflow of SLE

SLE streamlines sample preparation by combining LLE principles with a solid support, providing cleaner extracts and improved phase separation. This workflow is particularly valuable in forensic toxicology for isolating drugs, metabolites, and toxicants from complex biological matrices. The stepwise workflow of SLE is summarized in Fig. 6.18.

Key Steps

- **Sample Preparation**: Adjust sample pH (e.g., pH 9 for basic drugs), add salts such as NaCl to enhance partitioning, and dilute viscous samples to prevent clogging.
- **Conditioning (Optional)**: Pre-wet the solid support (e.g., diatomaceous earth) with water or buffer to enhance analyte retention.

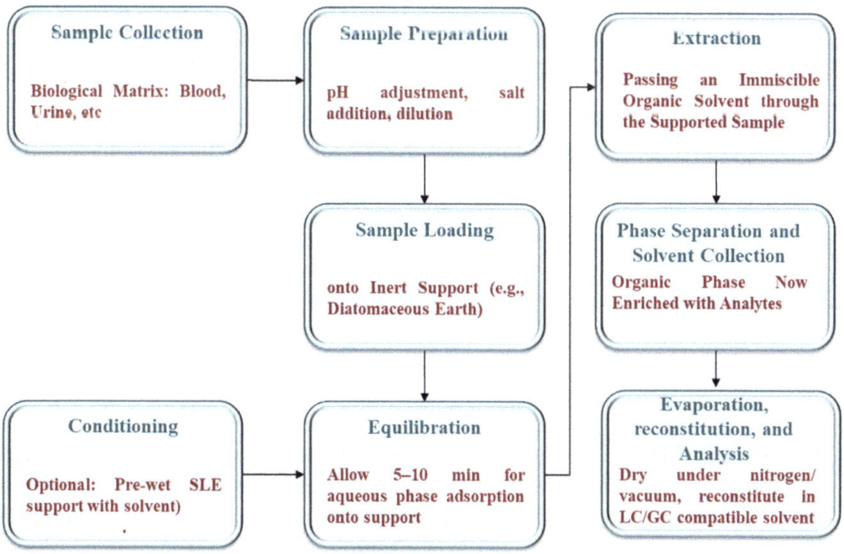

Fig. 6.18 Flow diagram illustrating the stepwise workflow of SLE

Table 6.22 Critical parameters in SLE method

Factor	Role
Solid support	High-surface-area material (e.g., diatomaceous earth) to retain aqueous phase
Organic solvent	Immiscible with water; selected based on analyte polarity (e.g., ethyl acetate for non-polar drugs)
pH adjustment	Optimizes ionization state (e.g., pH > pKa for basic drugs to remain uncharged)
Salt addition	Enhances partitioning via salting-out effect (e.g., NaCl or $MgSO_4$)

Table 6.23 Comparison of SLE with LLE

Parameter	SLE	LLE
Emulsion risk	None	High
Solvent use	0.5–2 mL	5–20 mL
Throughput	High (automated)	Low (manual)
Matrix clean up	Excellent	Moderate

- **Sample Loading**: Introduce the aqueous sample onto the SLE support, such as a cartridge or 96-well plate.
- **Equilibration**: Allow 5–10 min for the aqueous phase to adsorb, forming a stable interface.
- **Solvent Elution**: Apply an immiscible organic solvent to extract analytes.
- **Phase Collection**: Collect the analyte-rich organic phase, leaving matrix interferents behind.
- **Evaporation/Reconstitution**: Concentrate the extract if needed.
- **Analysis**: Analyze via LC–MS for polar compounds or GC–MS for volatiles.

Key factors influencing SLE efficiency are summarized in Table 6.22.

A comparative overview of SLE and traditional LLE methods is presented in Table 6.23.

6.8.4 Forensic Toxicological Applications

SLE has become a reliable tool in forensic toxicology for isolating and concentrating analytes from complex biological matrices.

- **Extraction of Drugs of Abuse**: SLE effectively isolates a wide range of drugs, including opioids, amphetamines, cocaine, and benzodiazepines, from blood, urine, and other biological fluids. Its robust extraction minimizes co-extraction of interfering compounds, ensuring cleaner samples for analysis.

- **Metabolite Analysis**: For confirming drug exposure or estimating the timing of intake, SLE pre-concentrates metabolites, enhancing their detectability with advanced analytical techniques such as LC–MS or GC–MS.
- **Toxicant and Pesticide Screening**: In cases of suspected poisoning or environmental exposure, SLE efficiently extracts toxicants and pesticide residues while reducing interference from endogenous matrix components, improving the accuracy and reliability of identification and quantification.

The primary forensic applications of SLE are summarized in Table 6.24.

Table 6.24 Key forensic toxicology applications of SLE [58–61]

Matrix	Analytes/drug class	SLE format & support	Extraction solvent	Performance
Urine	24 Drugs of Abuse (amphetamines, opioids)	Chem Elut S 400 mg 96-well plate	MTBE	Recoveries: 80–120% (±20%), RSD < 20%
Whole blood	Cocaine, Opiates, THC & Metabolites	ISOLUTE SLE + 1 mL column	MTBE	Recoveries: ≥85%, protein-free extract
Oral fluid	47 Drugs of Abuse	Chem Elut S 96-well plate (400 mg)	MTBE	Recoveries: 70–110%
Serum	Steroid Panel (cortisol, testosterone)	ISOLUTE SLE + 1 mL 48-well plate	MTBE/Hexane mixtures	Recoveries: 88–102%
Whole blood	Synthetic Cannabinoids (NBOMe analogs)	ISOLUTE SLE + 1 mL column	MTBE	Recoveries: 75–95%
Hair	Amphetamines and Analogues	ISOLUTE SLE + columns & plates	MTBE/Ethyl Acetate	Recoveries: 80–115%
Whole blood	Multi-class Drug Panel (>35 analytes)	Chem Elut S cartridge	MTBE	Limits of detection: 1–10 ng/mL
Postmortem samples	Opiates, Benzodiazepines	Diatomaceous Earth SLE cartridge	MTBE	Recoveries: 80–98%
Bile and tissues	Drugs of Abuse and Metabolites	ISOLUTE SLE + 2 mL column	MTBE/ Dichloromethane	Clean, phospholipid-free extracts
Plasma	Therapeutic Drugs (antidepressants)	SLE + 96-well plate	MTBE	Recoveries: 85–105%, RSD <15%

6.8.5 *Advantages and Limitations of Supported Liquid Extraction (SLE)*

Supported Liquid Extraction (SLE) offers a streamlined and efficient approach for isolating analytes from complex biological matrices, making it a valuable tool in forensic toxicology.

Advantages

- Reduced Emulsion Formation: The use of a solid support minimizes the formation of emulsions, a common issue in traditional LLE, resulting in cleaner and faster phase separation.
- High Throughput and Reproducibility: SLE can be readily automated, allowing laboratories to process large numbers of forensic samples quickly with minimal manual handling, enhancing consistency and reproducibility.
- Enhanced Selectivity: By adjusting solvent systems, pH, and other parameters, SLE can be tailored to selectively extract a wide range of analytes, including both polar and non-polar compounds.
- Cost Efficiency and Reduced Solvent Use: Smaller volumes of organic solvents are required compared to conventional LLE, lowering operational costs and minimizing environmental impact.

Limitations

- Dependence on Analyte Solubility: Extraction efficiency relies heavily on the partition coefficient of the analytes. Compounds with poor affinity for the chosen solvent may show reduced recovery.
- Method Development Needs: Optimization of parameters such as solvent type, solvent-to-sample ratio, and pH is often necessary for each new analyte or matrix.
- Potential Matrix Effects: Despite improved separation, complex biological samples may still introduce interfering substances, requiring additional cleanup or calibration steps.

Overall, SLE combines efficiency, adaptability, and environmental benefits, but careful optimization is essential to overcome limitations and ensure accurate forensic analysis.

6.9 Magnetic Solid-Phase Extraction (MSPE)

Magnetic Solid-Phase Extraction (MSPE) is an innovative sorption-based technique that uses magnetic nanoparticles (MNPs) to efficiently isolate and preconcentrate target analytes from complex sample matrices [62]. The approach relies on functionalized magnetic sorbents that interact selectively with analytes, followed by rapid separation via an external magnetic field. By dispersing MNPs into the sample, analytes adsorb onto their surfaces, and a magnet is then used to retrieve the loaded particles, leaving behind matrix interferences. The analytes are subsequently released using a small volume of elution solvent for instrumental analysis [63]. This technique eliminates the need for packed columns, centrifugation, or filtration, enhancing extraction speed, selectivity, and environmental sustainability.

The term MSPE was first introduced by Šafaříková and Šafařík in 1999, who used silanized magnetite particles to preconcentrate organic dyes [64]. Since then, a variety of magnetic sorbents, including polymer-coated magnetite, graphene oxide, and carbon-coated Fe_3O_4, have expanded MSPE's applications in forensic and environmental analysis.

6.9.1 Theory and Principle of MSPE

MSPE utilizes magnetic sorbents, typically Fe_3O_4 or γ-Fe_2O_3 nanoparticles, coated with materials such as polydopamine, silica, or metal–organic frameworks (MOFs). These coatings enhance selectivity through hydrophobic, ionic, or π-π interactions with analytes [62]. When dispersed in an aqueous sample, the sorbents provide a high surface area for adsorption. Analytes partition onto the MNPs according to their partition coefficients and specific chemical interactions. After adsorption equilibrium, usually reached within minutes, the nanoparticles are rapidly collected using an external magnet. The analytes are then eluted using solvents like methanol or acetonitrile, often with modifiers, for subsequent analysis. A schematic overview of the MSPE procedure is presented in Fig. 6.19.

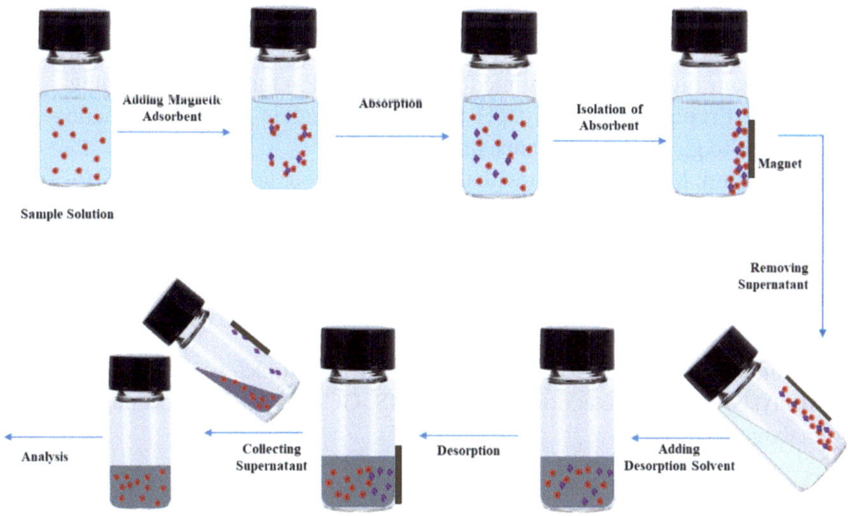

Fig. 6.19 Schematic representation of MSPE

6.9.2 Core Synthesis and Coating Strategies

Magnetite (Fe_3O_4) nanoparticles are commonly prepared via co-precipitation of Fe^{2+}/Fe^{3+} salts under alkaline conditions, with particle sizes controlled between 10 and 100 nm to ensure superparamagnetic behavior. Surface coatings tailor the sorbents' selectivity and functionality:

- Silica shells provide stability and reactive sites for functionalization.
- Polymeric layers (e.g., C18 or ion-exchange groups) enhance hydrophobic or ionic interactions.
- Carbon/Graphene coatings improve adsorption of aromatic and polar compounds.
- Molecularly Imprinted Polymers (MIPs) offer highly selective template-based binding.

Characterization of MSPE sorbents is performed using TEM, SEM, FTIR, VSM, and BET surface area analysis to verify morphology, magnetic properties, and surface functionality.

6.9.3 Workflow of MSPE

MSPE offers a rapid and efficient approach to isolate analytes from complex matrices by combining magnetic sorbents with selective chemical interactions. The workflow integrates sample preparation, adsorption, magnetic separation, and elution into a streamlined procedure suitable for forensic and environmental analyses.

- **Adsorbent Preparation**: Functionalize magnetic nanoparticles (Fe_3O_4) with coatings such as C18 for non-polar compounds, graphene oxide for aromatic analytes, or molecularly imprinted polymers (MIPs) for selective binding.
- **Sample Preparation**: Adjust pH to optimize analyte–sorbent interactions and add salts (e.g., NaCl) to enhance binding.
- **Adsorption**: Mix the sorbent with the sample (blood, urine, water) for 10–30 min to allow analyte capture.
- **Magnetic Separation**: Apply an external magnet to quickly isolate the analyte-loaded particles.
- **Washing**: Remove matrix interferences using mild solvents.
- **Elution**: Release analytes with an appropriate solvent.
- **Adsorbent Recycling**: Regenerate sorbents for multiple cycles.
- **Analysis**: Perform detection using LC–MS, GC–MS, or ICP-MS depending on analyte type.

The detailed workflow of MSPE is illustrated in Fig. 6.20.

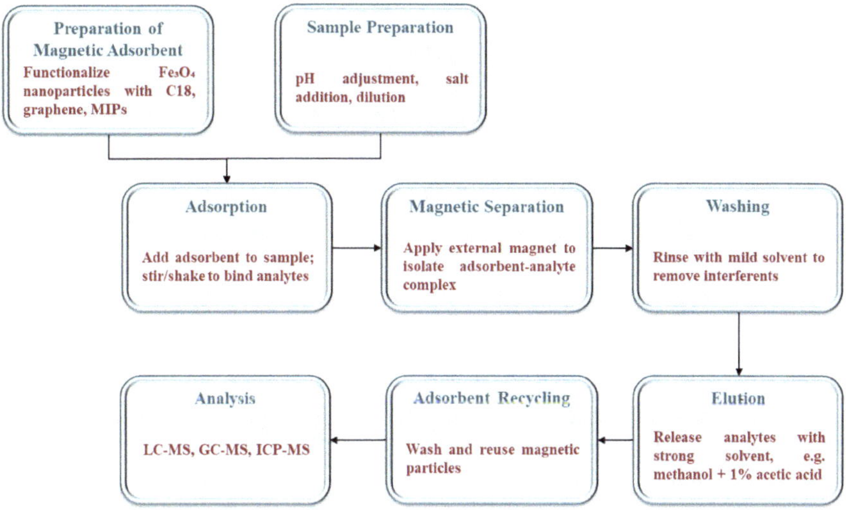

Fig. 6.20 Flow diagram that outlines the key steps in MSPE

6.9.4 Applications in Forensic Toxicology

Magnetic Solid-Phase Extraction (MSPE) has emerged as a highly effective tool in forensic toxicology for isolating and preconcentrating a broad spectrum of analytes from complex biological matrices such as blood and urine [64–68]. It has been successfully applied to detect antidepressants, antihypertensives, opioids, stimulants, and other drugs, offering high recovery rates, low detection limits, and compatibility with analytical platforms like LC–MS/MS and HPLC–UV.

- **Antidepressants**: C18-functionalized $Fe_3O_4@SiO_2$ nanoparticles efficiently extracted venlafaxine, paroxetine, fluoxetine, norfluoxetine, and sertraline from plasma and urine, with recoveries of 77–119% and CVs < 10.5%.
- **Antihypertensives**: Carbon-coated Fe_3O_4 nanoparticles measured losartan, carvedilol, and amlodipine in plasma, achieving recoveries of 62–97%.
- **Drugs of Abuse**: Mixed-mode MNPs combining ion-exchange and hydrophobic ligands enabled rapid isolation of amphetamines, opioids, and benzodiazepines from biological fluids for confirmatory LC–MS/MS analysis.
- **Emerging Contaminants**: Graphene oxide/Fe_3O_4 and C_{18}-Fe_3O_4 MNPs preconcentrated pharmaceuticals and artificial sweeteners from wastewater, achieving recoveries of 58–103% and detection limits as low as 1.1 ng L^{-1}.

The forensic applications of MSPE, including extraction of antidepressants, antihypertensives, drugs of abuse, and emerging contaminants, are summarized in Table 6.25.

Table 6.25 Forensic toxicological applications of MSPE

Application area	Target analytes	Sample type	MSPE sorbent used	Key advantages	Performance metrics
Illicit drug detection	Opioids, benzodiazepines, amphetamines	Blood, urine, hair	Fe_3O_4@polydopamine nanocomposites	High selectivity for polar drugs; removes matrix interferents (e.g., proteins, lipids)	Recovery: 85–95%; LOD: 0.1–5 ng/mL; Extraction time: 10–15 min
Pesticide analysis	Organophosphates, carbamates	Environmental water, soil	Cu-MOF@Fe_3O_4	Efficient adsorption of hydrophobic pesticides; reusable for 5–10 cycles	Recovery: >90%; LOD: 0.5–2 μg/L; RSD: <8% (n = 5)
Postmortem toxicology	Drugs, metabolites (e.g., fentanyl)	Decomposed tissues, vitreous humor	Fe_3O_4@SiO_2-C18	Tolerates putrefied matrices; minimizes co-extraction of fatty acids	Recovery: 75–88%; LOD: 10–50 pg/g; Matrix effect: <15%
Multi-analyte screening	Opioids, cannabinoids, antidepressants	Oral fluid, dried blood spots	Graphene-coated MNPs	Simultaneous extraction of diverse drug classes; compatible with LC–MS/MS	Recovery: 80–92%; Throughput: 20 samples/hour; Cross-reactivity: <5%
Heavy metal detection	Lead, arsenic, mercury	Hair, nails	Fe_3O_4@thiol-functionalized silica	Selective chelation of metals; eliminates organic interferents	Recovery: 88–95%; LOD: 0.01–0.1 μg/g; Precision: RSD <10%
Novel psychoactive substances (NPS)	Synthetic cathinones, cannabinoids	Wastewater, seized powders	Aptamer-functionalized MNPs	Captures low-abundance NPS; avoids antibody cross-reactivity	Recovery: 70–85%; LOD: 0.05–0.2 ng/mL; Specificity: >98%

6.9.5 Advantages of MSPE

MSPE offers a fast, efficient, and environmentally friendly approach for isolating and concentrating analytes from complex matrices. Its unique use of magnetic nanoparticles streamlines sample preparation while maintaining high sensitivity and selectivity, making it particularly valuable in forensic and environmental analyses.

- **Rapid and Simple Workflow**: Direct contact between sample and sorbent occurs in a single dispersive step, and magnetic retrieval is instantaneous, reducing total extraction time to minutes.
- **Minimal Solvent and Reagent Use**: Only 100–500 μL of elution solvent is needed, supporting green chemistry and reducing disposal costs.
- **High Enrichment and Sensitivity**: Functional coatings such as C_{18}, ion exchange, or molecularly imprinted polymers allow selective binding, yielding enrichment factors often > 50× and detection limits in the low ng mL^{-1} range.
- **Excellent Selectivity**: Tailored surface chemistries enable recovery of specific drug classes with high precision (77–119% recoveries, CV < 10%).
- **Automation Compatibility**: Magnetic separation integrates easily with robotic platforms, enhancing throughput and reproducibility.
- **Wide Applicability**: MSPE can be applied to blood, urine, wastewater, and food matrices, handling drugs of abuse, environmental contaminants, and emerging pollutants effectively.

The performance of MSPE compared to conventional SPE is summarized in Tables 6.26 and 6.27 highlights the major benefits of MSPE relative to SPE.

Table 6.26 Comparison of MSPE with traditional SPE

Parameter	MSPE	SPE
Separation time	1–5 min (magnet)	5–15 min (centrifugation/vacuum)
Adsorbent cost	Higher (nanoparticle synthesis)	Lower (pre-packed cartridges)
Reusability	Yes (5–10 cycles)	No (single-use cartridges)
Throughput	High (batch processing)	Moderate (column-based)

Table 6.27 Key advantages of MSPE over SPE

Aspect	MSPE	Traditional SPE
Separation	Instant (magnet)	Slow (vacuum/centrifugation)
Matrix tolerance	High (handles viscous samples like blood)	Moderate (prone to clogging)
Cost efficiency	Reusable adsorbents	Single-use cartridges

6.9.6 Limitation of MSPE

Magnetic Solid-Phase Extraction (MSPE) is a powerful and environmentally friendly technique, yet it has certain limitations that must be considered for effective application.

- **Nanoparticle Aggregation**: At higher sorbent amounts, magnetic nanoparticles (MNPs) can agglomerate, reducing surface area and extraction efficiency, unless stabilized coatings or sonication are used.
- **Variable Recoveries and Matrix Effects**: Differences in analyte affinity, sample composition, and pH can cause inconsistent recoveries, necessitating thorough method validation and internal standards.
- **Limited Sorbent Reusability**: Repeated use may degrade functional coatings or accumulate matrix contaminants, decreasing performance over successive cycles.
- **Optimization Requirements**: Parameters such as MNP functionalization, sorbent quantity, adsorption time, wash solvents, and elution conditions must be carefully tuned for each analyte.
- **Potential Leaching**: Inadequate coatings may release iron ions or other core materials, potentially affecting downstream analysis.
- **Synthesis Costs**: Initial preparation and functionalization of MNPs (e.g., silica, polymers, graphene, MIPs) can be complex and resource-intensive.

Despite these challenges, ongoing advancements in nanomaterial design continue to improve MSPE's robustness, selectivity, and applicability, making it a valuable tool for forensic and analytical sciences.

6.10 Dilute-And-Inject (D&I)

The dilute-and-inject (D&I) technique, also referred to as "dilute-and-shoot," is a simplified and rapid sample preparation method widely used in forensic toxicology. It is particularly effective for analyzing protein-poor biological fluids, including urine and saliva, where extensive cleanup is not required. The minimal sample handling involved makes D&I ideal for high-throughput workflows, allowing laboratories to process large numbers of samples efficiently while reducing preparation time and potential errors [23, 24]. As illustrated in Fig. 6.21, the D&I approach allows direct sample introduction into LC–MS following minimal dilution, simplifying analysis and reducing matrix interferences.

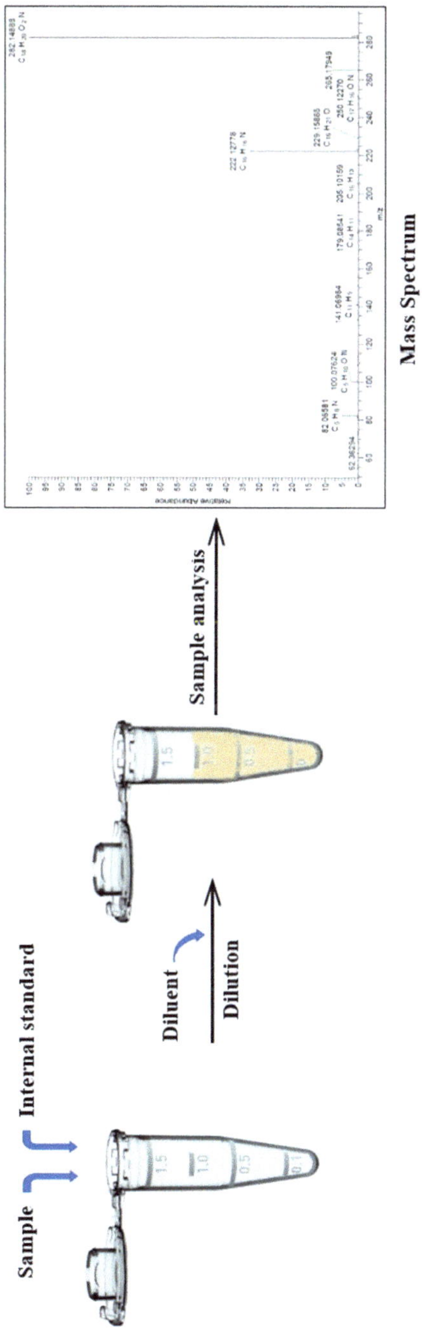

Fig. 6.21 Schematic illustration of the D&I method

6.10.1 Theory and Principle

In D&I approach, the biological sample is diluted using an appropriate solvent, often water or a solution compatible with the LC–MS mobile phase. Dilution serves to reduce matrix effects, lower ion suppression, and minimize potential interferences during instrumental analysis [69]. Following dilution, the sample is directly injected into the LC–MS system without additional extraction or cleanup steps. Because protein content is low in these matrices, the risk of column clogging or instrumental fouling is significantly decreased. This streamlined workflow maintains analytical sensitivity and reproducibility while simplifying sample preparation, making D&I a practical choice for rapid screening and routine forensic investigations.

6.10.2 Workflow of D&I

The D&I method streamlines sample preparation by minimizing handling while maintaining analytical reliability.

- **Sample Collection**: Obtain the biological specimen, such as urine, plasma, or saliva.
- Addition of Internal Standard: Introduce a known quantity of an internal standard to improve quantification and compensate for variability during analysis.
- **Dilution with Solvent**: Dilute the sample using an appropriate solvent, often water or a buffer compatible with the LC–MS system, to reduce matrix effects and potential interferences.
- **Mixing**: Ensure thorough homogenization of the diluted sample.
- **Optional Filtration or Centrifugation**: Remove particulates if necessary to prevent LC–MS system blockage.
- **Transfer to Vial**: Place the prepared sample into an auto-sampler vial suitable for the instrument.
- **Direct Injection**: Inject the diluted sample directly into the LC–MS system for analysis.

As shown in Fig. 6.22, the D&I method simplifies sample preparation by combining dilution and direct injection for efficient LC–MS analysis.

6.10.3 Applications in Forensic Toxicology

The D&I approach has gained widespread use in forensic toxicology due to its simplicity and efficiency in detecting and quantifying various compounds [70–72]. It is particularly effective for:

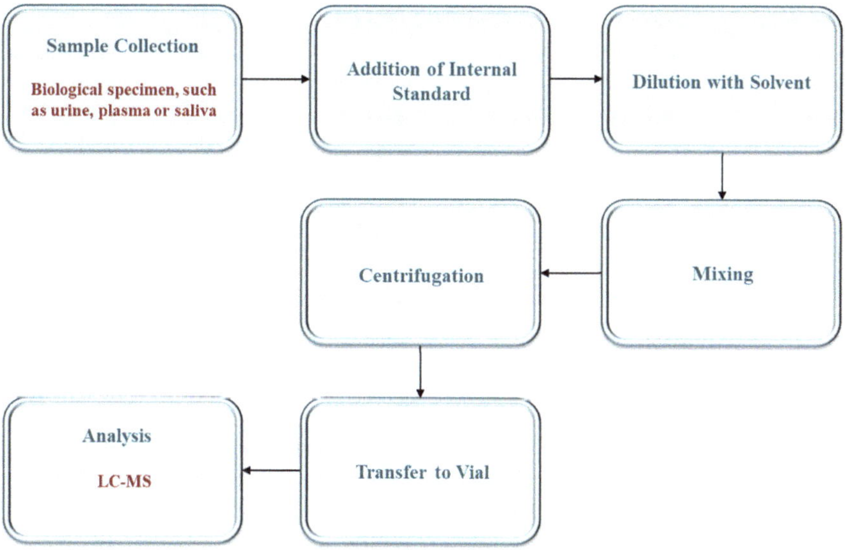

Fig. 6.22 Flow diagram illustrating the key steps in the dilute-and-inject (D&I) workflow

- **Drugs of Abuse**: Including amphetamines, opioids, and benzodiazepines.
- **Prescription Medications**: Such as antihypertensives, antipsychotics, and other commonly prescribed drugs.
- **Doping Agents**: Applied in sports medicine for urine drug screening and anti-doping programs.

By coupling D&I with high-resolution mass spectrometry, laboratories can perform comprehensive screening of complex biological matrices with minimal sample preparation, ensuring rapid and reliable results.

Primary Target Matrices

- *Urine*: Favored for its higher analyte concentrations and low protein content.
- *Oral Fluid*: Offers non-invasive collection and is increasingly used in roadside drug testing scenarios.

As summarized in Table 6.28, the D&I method enables rapid and efficient analysis of drugs, prescription medications, and doping agents across multiple biological matrices.

Table 6.29 presents a comparative overview of the D&I method and conventional extraction techniques such as LLE and SPE. The comparison emphasizes the operational simplicity, reduced solvent consumption, and enhanced throughput of D&me, highlighting its suitability for high-efficiency forensic toxicological applications where rapid and reliable sample processing is essential.

Table 6.28 Forensic toxicological applications of D&I method

Application area	Target analytes	Sample type	Key advantages	Limitations	Performance metrics
Drug screening in urine	Amphetamines, opioids, cannabinoids	Urine	Minimal sample prep (1:10 dilution); compatible with LC–MS/MS and GC–MS	Matrix effects (e.g., salts, urea) may suppress ionization	Recovery: 85–95%; LOD: 1–5 ng/mL; Throughput: 100 + samples/day
Blood alcohol analysis	Ethanol, methanol, GHB	Whole blood, serum	Rapid detection (≤5 min/sample); avoids derivatization steps	Limited sensitivity for trace volatiles (e.g., acetone)	Recovery: 90–98%; LOQ: 10 mg/dL; RSD: <5%
Oral fluid drug testing	Cocaine, MDMA, benzodiazepines	Oral fluid (saliva)	Non-invasive; ideal for roadside or workplace testing	Low analyte concentrations require high-sensitivity instruments	Recovery: 70–88%; LOD: 0.5–2 ng/mL; Matrix effect: 15–25%
Postmortem toxicology	Antidepressants, antipsychotics	Vitreous humor, decomposed blood	Bypasses protein precipitation; suitable for unstable matrices	Co-eluting interferents (e.g., putrefactive amines) reduce accuracy	Recovery: 60–80%; LOD: 5–10 ng/mL; Carryover: <0.1%
Novel psychoactive substances (NPS)	Synthetic cathinones, cannabinoids	Seized powders, diluted blood	Detects polar/non-polar NPS without hydrolysis or extraction	Poor selectivity for structural analogs (e.g., fentanyl derivatives)	Recovery: 50–75%; LOD: 0.2–1 ng/mL; False positives: 2–5%
Emergency toxicology	Acetaminophen, salicylates	Plasma, serum	Rapid turnaround (<10 min) for overdose cases; no solvent waste	High lipid content in plasma causes column fouling	Recovery: 80–90%; LOQ: 50 ng/mL; Throughput: 50 samples/hour

Table 6.29 Comparison of D&I with traditional extraction methods

Parameter	Dilute-and-inject	Solid-phase extraction (SPE)	Liquid–liquid extraction (LLE)
Prep time	5–10 min	30–60 min	45–90 min
Solvent use	≤100 μL/sample	5–10 mL/sample	10–20 mL/sample
Cost/sample	0.50–1.00	2.00–5.00	3.00–8.00
Sensitivity	Moderate (ng/mL)	High (pg/mL)	High (pg/mL)
Matrix tolerance	Moderate (requires dilution)	High (removes interferents)	Low (emulsion risks)

6.10.4 Advantages and Limitations of the Dilute-And-Inject (D&I) Method

The dilute-and-inject technique offers a fast, simple, and cost-efficient approach for forensic and analytical applications. By minimizing sample preparation steps, it enables high-throughput analysis while reducing solvent use and handling errors. However, its performance can be influenced by matrix complexity and instrument sensitivity requirements.

Advantages

- *Simplicity and Speed*: Eliminates extensive extraction steps, allowing rapid and straightforward analysis.
- *Cost Effectiveness*: Requires minimal solvents and consumables, reducing overall operational costs.
- *High Throughput*: Enables efficient processing of multiple samples in routine forensic screening.
- *Reduced Sample Handling*: Lowers the risk of contamination and analyte loss, enhancing reproducibility.

Limitations

- *Matrix Effects*: Interfering components in the sample matrix may impact analytical precision.
- *Sensitivity Constraints*: Detecting trace analytes can be difficult without pre-concentration steps.
- *Instrument Maintenance*: Frequent direct injections can cause contamination,

The D&I technique serves as a fast, economical, and practical approach for large-scale drug screening, particularly effective for urine and oral fluid analysis in forensic toxicology. Despite certain limitations in sensitivity and matrix management, continuous advancements in high-resolution mass spectrometry and automation are enhancing its precision and expanding its role in detecting emerging substances such as novel psychoactive compounds.

6.11 Summary

In this chapter, the diverse extraction techniques central to forensic toxicology and anti-doping studies have been critically examined, emphasizing their operational principles, performance advantages, and methodological limitations. Each approach contributes uniquely to the accurate identification of drugs and doping agents, highlighting the importance of tailored method selection in analytical practice.

LLE continues to be valued for its simplicity, though its reliance on large solvent volumes and limited selectivity restricts its use in complex cases. SPE and SPME demonstrate significant advancements, offering reproducible, efficient, and environmentally sustainable alternatives. Miniaturized innovations such as DLLME and SBSE provide excellent sensitivity and reduced solvent use, aligning with modern demands for greener and more efficient workflows. Similarly, SLE and MSPE reflect the trend toward automation, reusability, and enhanced selectivity, thereby supporting large-scale forensic and doping investigations. The dilute-and-inject approach, while appealing for its speed, remains confined to relatively clean samples and requires robust instrumentation to ensure reliability.

The comparison of these methods underscores a fundamental principle: no single extraction technique meets all analytical needs. Instead, effective toxicological analysis depends on balancing sensitivity, selectivity, cost, and practicality in accordance with the case or testing requirements. For anti-doping laboratories, this means aligning extraction strategies with the WADA stringent standards, while forensic laboratories must ensure defensible and reproducible results in legal contexts.

Ultimately, this chapter concludes that continued innovation in extraction science through miniaturization, green chemistry, and automation will enhance both forensic and anti-doping research. By integrating classical and emerging methods, laboratories can achieve greater analytical precision, protect public health, and uphold fairness in sport, ensuring that scientific progress translates into ethical and reliable practice.

References

Jones S, McGowan C, Boyle S, et al. An overview of sample preparation in forensic toxicology. Wiley Interdiscip Rev Forensic Sci. 2022;4: e1436.

Celestin MN, Musteata FM. Impact of changes in free concentrations and drug-protein binding on drug dosing regimens in special populations and disease states. J Pharm Sci. 2021;110:3331–44.

Olariu RI, Vione D, Grinberg N, et al. Sample preparation for trace analysis by chromatographic methods. J Liq Chromatogr Relat Technol. 2010;33:1174–207. https://doi.org/10.1080/10826076.2010.484371.

TDSSA. Technical document for sport specific analysis. https://www.wada-ama.org/sites/default/files/2025-01/tdssa_v10.0_final_en-2.pdf (2010).

Qallaf MA. Different extraction procedures for multiple drugs abuse in human hair: case study. Chem Pharm Res. 2022;4:1–6.

Ingle RG, Zeng S, Jiang H, et al. Current developments of bioanalytical sample preparation techniques in pharmaceuticals. J Pharm Anal. 2022;12:517–29.

Olariu RI, Vione D, Grinberg N, Arsene C. Sample preparation for trace analysis by chromatographic methods. J Liq Chromatogr Relat Technol. 2010;33:1174–207.
Rygaard K, Pan M, Nielsen MKK, et al. Overview of systematic toxicological analysis strategies and their coverage of substances in forensic toxicology. Anal Sci Adv. 2023;4:96–103.
USFDA-Bioanalytical Method Validation Guidance for Industry. https://www.fda.gov/files/drugs/published/Bioanalytical-Method-Validation-Guidance-for-Industry.pdf (2018).
ICH guideline M10 on bioanalytical method validation and study sample analysis. https://www.ema.europa.eu/en/documents/scientific-guideline/guideline-bioanalytical-method-validation_en.pdf (2022).
Bakarr KA. Recent developments in sample preparation techniques combined with high-performance liquid chromatography: a critical review. J Chromatogr A. 2021;1654: 462444. https://doi.org/10.1016/j.chroma.2021.462444.
Ma L, Yang L, Tang L, et al. Sample preparation techniques for quality evaluation and safety control of medicinal and edible plants: overview advances, applications, and future perspectives. J Pharm Anal. 2025;101296. https://doi.org/10.1016/j.jpha.2025.101296
Frederick F, Cantwell ML. Chapter 11: Liquid-liquid extraction. In: Comprehensive Analytical Chemistry, Elsevier, Vol 37, pp. 297–340 (2002). https://doi.org/10.1016/S0166-526X(02)80048-4
Badawy MEI, El-Nouby MAM, Kimani PK, et al. A review of the modern principles and applications of solid-phase extraction techniques in chromatographic analysis. Anal Sci. 2022;38:1457–87. https://doi.org/10.1007/s44211-022-00190-8.
Liu S, Ouyang G. Introduction of solid-phase microextraction. In: Ouyang G, Jiang R (eds) Solid Phase Microextraction. Springer, Berlin, Heidelberg (2017). https://doi.org/10.1007/978-3-662-53598-1_1
Zgoła-Grześkowiak A, Grześkowiak T. Dispersive liquid-liquid microextraction. TrAC Trends Anal Chem. 2011;30:1382–99. https://doi.org/10.1016/j.trac.2011.04.014.
Pereira J, Gonçalves J, Alves V, et al. Microextraction using packed sorbent as an effective and high-throughput sample extraction technique: recent applications and future trends. Sample Preparation, 38–53 (2013). https://doi.org/10.2478/sampre-2013-0005
Moein MM, Said R, Abdel-Rehim M. Microextraction by packed sorbent. Bioanalysis. 2015;7:2155–61. https://doi.org/10.4155/bio.15.154.
David F, Sandra P. Stir bar sorptive extraction for trace analysis. J Chromatogr A. 2007;1152:54–69. https://doi.org/10.1016/j.chroma.2007.01.032.
He M, Wang Y, Zhang Q, et al. Stir bar sorptive extraction and its application. J Chromatogr A. 2021;1637: 461810. https://doi.org/10.1016/j.chroma.2020.461810.
Ronald E, Majors. Supported liquid extraction (SLE): the best kept secret in sample preparation. LCGC North Am. 2012;30(8):626–633.
Hagarová I. Magnetic solid phase extraction as a promising technique for fast separation of metallic nanoparticles and their ionic species: a review of recent advances. J Anal Methods Chem. 2020;11:8847565. https://doi.org/10.1155/2020/8847565.
Greer B, Chevallier O, Quinn B, et al. Redefining dilute and shoot: the evolution of the technique and its application in the analysis of foods and biological matrices by liquid chromatography mass spectrometry. TrAC Trends Anal Chem. 2021;141: 116284. https://doi.org/10.1016/j.trac.2021.116284.
Tamama K. Dilute and shoot approach for toxicology testing. Front Chem. 2023;11:1278313. https://doi.org/10.3389/fchem.1278313.
Zhang J, Hu B. Liquid-liquid extraction (LLE). In: Ramaswamy S, Huang HJ, Ramarao BV (eds) Separation and Purification Technologies in Biorefineries. Wiley, Hoboken (2013). https://doi.org/10.1002/9781118493441.ch3
Wells DA. Chapter 9-Liquid-liquid extraction: strategies for method development and optimization. Prog Pharm Biomed Anal. 2003;5:307–326. https://doi.org/10.1016/S1464-3456(03)80011-X
Taylor T. Strategies for optimization of liquid-liquid extraction (LLE) protocols. LCGC North Am. 2021;39(4):198.

Buszewski B, Szultka M. Past, present, and future of solid phase extraction: a review. Crit Rev Anal Chem. 2012;42:198–213. https://doi.org/10.1080/07373937.2011.645413.
Hennion MC. Solid-phase extraction: method development, sorbents, and coupling with liquid chromatography. J Chromatogr A. 1999;856:3–54. https://doi.org/10.1016/s0021-9673(99)00832-8.
Thurman EM, Mills MS. Solid-phase extraction: principles and practice. New York: Wiley-Interscience; 1998.
Raynie DE, Watson DW. Understanding and improving solid-phase extraction. LCGC North Am. 2014;32(12):908–15.
Mahdavijalal M, Petio C, Staffilano G, et al. Innovative solid-phase extraction strategies for improving the advanced chromatographic determination of drugs in challenging biological samples. Molecules. 2024;29:2278. https://doi.org/10.3390/molecules29102278.
Risticevic S, Lord H, Górecki T, et al. Protocol for solid-phase microextraction method development. Nat Protoc. 2010;5:122–39. https://doi.org/10.1038/nprot.2009.179.
Reyes-Garcés N, Gionfriddo E, Gómez-Ríos GA, et al. Advances in solid phase microextraction and perspective on future directions. Anal Chem. 2018;90:302–60. https://doi.org/10.1021/acs.analchem.7b04502.
Vékey K. Solid-phase microextraction: a powerful sample preparation tool prior to mass spectrometric analysis. J Mass Spectrom. 2004;39:233–54. https://doi.org/10.1002/jms.606.
De Giovanni N, Marchetti D. A systematic review of solid-phase microextraction applications in the forensic context. J Anal Toxicol. 2020;44:268–97. https://doi.org/10.1093/jat/bkz077.
Kumazawa T, Lee XP, Sato K, et al. Solid-phase microextraction and liquid chromatography/mass spectrometry in drug analysis. Anal Chim Acta. 2003;492:49–67. https://doi.org/10.1016/S0003-2670(03)00680-9.
Wells DA. Chapter 9 liquid-liquid extraction: strategies for method development and optimization. Prog Pharm Biomed Anal. 2003;5:307–326. https://doi.org/10.1016/S1464-3456(03)80011-X
Zuloaga O, Olivares M, Navarro P, et al. Dispersive liquid-liquid microextraction: trends in the analysis of biological samples. Bioanalysis. 2015;7:2211–25. https://doi.org/10.4155/bio.15.141.
Quigley A, Cummins W, Connolly D. Dispersive liquid-liquid microextraction in the analysis of milk and dairy products: a review. J Chem. 2016;2016:4040165. https://doi.org/10.1155/2016/4040165.
Zang XH, Wu QH, Zhang MY, et al. Developments of dispersive liquid-liquid microextraction technique. Chin J Anal Chem. 2009;37:161–8. https://doi.org/10.1016/S1872-2040(08)60082-1.
Moein MM, Abdel-Rehim A, Abdel-Rehim M. Microextraction by packed sorbent (MEPS). TrAC Trends Anal Chem. 2015;67:34–44. https://doi.org/10.1016/j.trac.2014.12.003.
Abdel-Rehim M. Microextraction by packed sorbent (MEPS): a tutorial. Anal Chim Acta. 2011;701:119–28. https://doi.org/10.1016/j.aca.2011.05.037.
Martins RO, Borsatto JVB, Will C, et al. Advancements in microextraction by packed sorbent: insights into sorbent phases and automation strategies. Separations. 2025;12:11. https://doi.org/10.3390/separations12010011.
Abdel-Rehim M. Current advances in microextraction by packed sorbent (MEPS) for bioanalysis applications. LCGC Europe. 2009;22(1):8–19.
Abdel-Rehim M. Recent advances in microextraction by packed sorbent for bioanalysis. J Chromatogr A. 2010;1217:2569–80. https://doi.org/10.1016/j.chroma.2009.09.053.
Wietecha-Posłuszny R, Garbacik A, Woźniakiewicz M, et al. Application of microextraction by packed sorbent to isolation of psychotropic drugs from human serum. Anal Bioanal Chem. 2012;402:2249–57. https://doi.org/10.1007/s00216-011-5442-0.
Ares AM, Fernández P, Regenjo M, et al. A fast bioanalytical method based on microextraction by packed sorbent and UPLC-MS/MS for determining new psychoactive substances in oral fluid. Talanta. 2017;174:454–61. https://doi.org/10.1016/j.talanta.06.022.
Pereira J, Câmara JS, Cojmsjö A, et al. Microextraction by packed sorbent: an emerging, selective and high-throughput extraction technique in bioanalysis. Biomed Chromatogr. 2014;28:839–47. https://doi.org/10.1002/bmc.3156.

Baltussen E, Sandra P, David F, et al. Stir bar sorptive extraction (SBSE), a novel extraction technique for aqueous samples: theory and principles. J Chromatogr Micro. 1999;11:737–47.
Zuloaga O, Etxebarria N, González-Gaya B, et al. Stir-bar sorptive extraction. In: Poole CF (ed) Handb Sep Sci, Solid-Phase Extraction. Elsevier, Amsterdam, pp. 493–530 (2020). https://doi.org/10.1016/B978-0-12-816906-3.00018-2
David F, Sandra P. Stir bar sorptive extraction for trace analysis. J Chromatogr A. 2007;1152:54–69. https://doi.org/10.1016/j.chroma.01.032.
Frank D, Ochiai N, Pat S. Two decades of stir bar sorptive extraction: a retrospective and future outlook. TrAC Trends Anal Chem. 2019;112:102–11. https://doi.org/10.1016/j.trac.2018.12.006.
Ochiai N, Sasamoto K, Frank D, et al. Recent developments of stir bar sorptive extraction for food applications: extension to polar solutes. J Agric Food Chem. 2018;66:7249–55. https://doi.org/10.1021/acs.jafc.8b02182.
Camino-Sánchez FJ, Rodríguez-Gómez R, Zafra-Gómez A, et al. Stir bar sorptive extraction: recent applications, limitations and future trends. Talanta. 2014;130:388–99. https://doi.org/10.1016/j.talanta.2014.07.022.
Prieto A, Basauri O, Rodil R, et al. Stir-bar sorptive extraction: a view on method optimisation, novel applications, limitations and potential solutions. J Chromatogr A. 2010;1217:2642–66. https://doi.org/10.1016/j.chroma.2009.12.051.
Silvestro L, Rizea SS. An update on solid phase-supported liquid extraction. Bioanalysis. 2015;7:2177–86. https://doi.org/10.4155/bio.15.144.
Cheng Z, Jiang H. Supported liquid extraction (SLE) in LC-MS bioanalysis. In: Li W, Jian W, Fu Y (eds) Sample Preparation in LC-MS Bioanalysis. Wiley, Hoboken (2019). https://doi.org/10.1002/9781119274315.ch6
Zhang Y, Cao H, Jiang H. Supported liquid extraction versus liquid-liquid extraction for sample preparation in LC-MS/MS-based bioanalysis. Bioanalysis. 2013;5:285–8. https://doi.org/10.4155/bio.12.335.
Jones R, Williams L, Lamboley C, et al. Drugs of abuse extraction from whole blood using supported liquid extraction (SLE) and Extrahera automation prior to UPLC-MS/MS analysis. Toxicol Anal Clin. 2018;30(2):S62. https://doi.org/10.1016/j.toxac.2018.04.082.
Rositano J, Harpasm P, Kostakis C, et al. Supported liquid extraction (SLE) for the analysis of methylamphetamine, methylenedioxymethylamphetamine, and delta-9-tetrahydrocannabinol in oral fluid and blood of drivers. Forensic Sci Int. 2016;265:125–30. https://doi.org/10.1016/j.forsciint.2016.01.017.
Šafaříková M, Šafařík I. Magnetic solid-phase extraction. J Magn Magn Mater. 1999;194:108–12. https://doi.org/10.1016/S0304-8853(98)00566-6.
Vasconcelos I, Fernandes C. Magnetic solid phase extraction for determination of drugs in biological matrices. TrAC Trends Anal Chem. 2017;89:41–52. https://doi.org/10.1016/j.trac.2016.11.011.
Chen LX, Yang FQ. Applications of magnetic solid-phase extraction in the sample preparation of natural product analysis. J Sep Sci. 2024;47:2400082. https://doi.org/10.1002/jssc.202400082.
Capriotti AL, Cavaliere C, La Barbera G, et al. Recent applications of magnetic solid-phase extraction for sample preparation. Chromatographia. 2019;82:1251–74. https://doi.org/10.1007/s10337-019-03721-0.
Moon H, Park J, Hwang J, et al. Magnetic solid-phase extraction of drugs and pesticides from human plasma using COOH-mMWCNTs. J Anal Toxicol. 2020;44:968–75. https://doi.org/10.1093/jat/bkaa029.
Yuan S, Xiang Y, Chen L, et al. Magnetic solid-phase extraction based on polydopamine-coated magnetic nanoparticles for rapid and sensitive analysis of eleven illicit drugs and metabolites in wastewater with the aid of UHPLC-MS/MS. J Chromatogr A. 2024;1718: 464703. https://doi.org/10.1016/j.chroma.2024.464703.
Evrim S, Durmus KZ. Magnetic solid phase extraction applications combined with analytical methods for determination of drugs in different matrices review. J Chil Chem Soc. 2019;64:4448–58. https://doi.org/10.4067/S0717-97072019000204448.

Deventer K, Pozo OJ, Verstraete AG, et al. Dilute-and-shoot-liquid chromatography-mass spectrometry for urine analysis in doping control and analytical toxicology. TrAC Trends Anal Chem. 2014;55:1–13. https://doi.org/10.1016/j.trac.2013.10.012.
Arena A, Ferracane A, Zoccali M, et al. A dilute-and-inject liquid-gas chromatography-mass spectrometry method for the analysis of sixteen polycyclic aromatic hydrocarbons in extra-virgin olive oil. Anal Chim Acta. 2024;1312: 342740. https://doi.org/10.1016/j.aca.2024.342740.
Görgens C, Guddat S, Orlovius AK, et al. Dilute-and-inject' multi-target screening assay for highly polar doping agents using hydrophilic interaction liquid chromatography high resolution/high accuracy mass spectrometry for sports drug testing. Anal Bioanal Chem. 2015;407:5365–79. https://doi.org/10.1007/s00216-015-8699-x.
DosSantos BP, Birk L, Schwarz P, et al. Validated dilute-and-shoot LC-MS/MS urine screening for the analysis of 95 illicit drugs and medicines: insights from clinical and forensic Brazilian cases. J Anal Toxicol. 2024;48:314–31. https://doi.org/10.1093/jat/bkae005.

Chapter 7
Comparative Case Studies on Extraction Methods for LC–MS and GC–MS Detection in Forensic Doping Control

Abstract This chapter provides a comprehensive comparative overview of extraction techniques utilized in forensic doping control for the identification of various classes of prohibited substances across biological matrices, including urine, plasma, and hair. Emphasis is placed on optimizing sample preparation to enhance analytical precision and reliability in both LC–MS and GC–MS analyses. Depending on the physicochemical properties of the target analytes, solid-phase extraction (SPE) and liquid–liquid extraction (LLE) were employed. Antihistamines, NSAIDs, corticosteroids, and anabolic steroids demonstrated efficient extraction through SPE in urine and LLE in plasma samples. Peptide-based drugs, myo-inositol trispyrophosphate (ITPP), and bisphosphonates achieved optimal recovery using SPE for both matrices, while quaternary ammonium compounds were effectively processed via direct dilution and injection (D&I). Furthermore, specialized protocols for hair extraction are presented, emphasizing their role in assessing long-term drug exposure. The chapter also explores the extraction and derivatization of non-polar and poorly ionizable compounds, such as specific anabolic steroids, which require chemical modification into volatile derivatives to enable GC–MS detection. In addition, the discussion extends to essential validation parameters such as selectivity, sensitivity, linearity, accuracy, precision, recovery, and matrix effects evaluated according to internationally recognized standards, including SWGTOX, ISO/IEC 17,025, and WADA guidelines. Overall, this comparative analysis underscores the significance of matrix-specific extraction and validation strategies in achieving robust, reproducible, and legally defensible results in forensic doping investigations.

Keywords Analytical toxicology · Forensic doping control · Anti-doping analysis · Biological matrices · Matrix-specific extraction · Sample preparation and cleanup · Derivatization and drug detection · Method validation

A. K. Karakka Kal et al., *Advanced Extraction Techniques for Drug and Toxin Detection*, https://doi.org/10.1007/978-3-032-17335-5_7

7.1 Overview

Forensic doping control serves as a backbone of sports integrity, ensuring fair competition and athlete safety through the detection of prohibited substances. The accuracy of such testing largely depends on the sample preparation stage, where analytes are effectively isolated and purified from complex biological matrices prior to LC–MS or GC–MS analysis [1, 2]. SPE is widely preferred for compounds requiring high selectivity and cleaner extracts. The method utilizes specific sorbents (e.g., C_{18}, ion-exchange, or mixed-mode) to retain analytes such as anabolic steroids, beta-blockers, diuretics, and peptide hormones while eliminating matrix interferences. SPE is particularly suited for polar or semi-polar compounds in plasma and urine, providing excellent recovery rates and compatibility with LC–MS/MS and GC–MS systems [3–6].

LLE, on the other hand, is often chosen for non-polar or moderately polar substances such as stimulants, narcotics, cannabinoids, and corticosteroids [4, 6–10]. This technique relies on differential solubility between aqueous and organic phases, offering high enrichment efficiency. LLE remains valuable in urine and plasma analysis where a wide range of lipophilic drugs must be extracted with minimal matrix interference, despite its relatively higher solvent consumption and manual handling requirements.

In contrast, the D&I approach provides a simplified, rapid, and reagent-minimal option for samples like urine and oral fluids, which are naturally low in protein content [11]. This method is typically selected for screening volatile, water-soluble, or low-protein-binding compounds such as amphetamines, opioids, and other drugs of abuse. Though D&I offers lower sensitivity compared to extraction-based methods, its high throughput and minimal sample processing make it indispensable for initial or large-scale screening.

Overall, the choice of extraction method in forensic doping control depends on the physicochemical properties of the drug, matrix composition, and analytical objectives. Through the case studies presented in this chapter, the strengths, limitations, and analytical outcomes of SPE, LLE, and D&I are discussed, offering a comprehensive understanding of their role in accurate, reproducible, and efficient doping analysis.

7.2 Sample Preparation and Extraction Techniques for Drug Analysis by LC–MS

7.2.1 Antihistamines

Antihistamines are a diverse group of drugs that block histamine receptors, commonly used to treat allergies, gastric disorders, and motion sickness [12]. Chemically, most antihistamines are basic in nature, which influences their extraction and detection in biological samples. Their physicochemical properties, including polarity and ionization, make them suitable for extraction using solid-phase or liquid–liquid extraction techniques prior to LC–MS analysis. Understanding these characteristics is crucial for developing reliable, sensitive, and reproducible analytical methods in forensic and clinical toxicology. The molecular characteristics of the selected antihistamine drugs are summarized in Table 7.1.

Extraction Procedure for Antihistamine Drug in Urine (SPE Method)

Urine samples are first pre-treated to release the target analytes from their conjugated forms and to improve their compatibility with the SPE process [13]. This preparatory step is essential for removing interfering matrix components such as proteins, salts, and metabolites that could reduce extraction efficiency or compromise analytical accuracy. After pre-treatment, the samples are processed using SPE, which enables selective retention and elution of analytes while reducing manual handling and the potential for error [14, 15].

Extraction Procedure

Sample Pre-treatment

Sample preparation begins with accurately dispensing 5 mL of urine into clean glass beakers, followed by adjustment of the pH to 5.0 using an appropriate buffer to

Table 7.1 Molecular characteristics of selected antihistamine Drugs

Drug name	Molecular formula	Molecular weight (g/mol)
Diphenhydramine	$C_{17}H_{21}NO$	255.36
Chlorpheniramine	$C_{16}H_{19}ClN_2$	274.79
Clemastine	$C_{21}H_{26}ClN$	327.88
Hydroxyzine	$C_{21}H_{27}ClN_2O_2$	374.90
Brompheniramine	$C_{16}H_{19}BrN_2$	313.24
Promethazine	$C_{17}H_{20}N_2S$	284.42
Cyproheptadine	$C_{21}H_{21}N$	291.41
Loratadine	$C_{22}H_{23}ClN_2O_2$	382.88
Cetirizine	$C_{21}H_{25}ClN_2O_3$	388.89
Levocetirizine	$C_{21}H_{25}ClN_2O_3$	388.89
Fexofenadine	$C_{32}H_{39}NO_4$	502.66
Desloratadine	$C_{19}H_{19}ClN_2$	310.82
Ranitidine	$C_{13}H_{22}N_4O_3S$	314.41
Famotidine	$C_6H_{15}N_7O_2S_3$	337.41
Nizatidine	$C_{12}H_{21}N_5O_3S$	307.39
Cimetidine	$C_6H_{16}N_6S$	252.34
Pitolisant	$C_{16}H_{19}ClN_2O$	280.79

establish optimal conditions for enzymatic hydrolysis. An internal standard solution is added to each sample to support accurate quantification and to compensate for potential matrix effects during analysis. Enzymatic hydrolysis is carried out by adding trypsin and β-glucuronidase (Type HP-2, ≥100,000 units/mL), with gentle mixing to ensure uniform enzyme distribution. The samples are incubated either overnight at 37 °C or for 2 h at 56 °C to achieve complete cleavage of glucuronide and sulfate conjugates. After incubation, the samples are cooled to room temperature, the pH is adjusted to 6.0–6.5 using dilute ammonium hydroxide, and the mixtures are centrifuged to sediment the particles. The clear supernatants are then collected for solid-phase extraction (SPE).

Solid Phase Extraction

Solid-phase extraction (SPE) is performed using cation-exchange cartridges. The cartridges are conditioned with methanol, followed by phosphate buffer, to activate and equilibrate the sorbent. The prepared samples are diluted with 0.1 M potassium dihydrogen phosphate buffer (pH 6.0) and loaded at a controlled flow rate to ensure effective analyte retention. This is followed by sequential washing with 0.1 M potassium dihydrogen phosphate buffer and 0.1 M acetic acid to remove interferences. The cartridges are then dried under oxygen-free nitrogen, and the analytes are eluted using a mixture of ethyl acetate, 2-propanol, and ammonia (96:2:2), with the eluate collected in 8 mL round-bottom threaded culture tubes.

Post-Extraction Processing

Following extraction, the eluates are evaporated to dryness under a gentle nitrogen stream at temperatures below 60 °C. The residues are reconstituted in a methanol–water mixture (1:1), vortex-mixed, centrifuged, and the clear solutions are transferred to autosampler vials for subsequent LC–MS analysis.

Each analytical batch includes a biological blank, test samples, and a spiked control for quality assurance.

Extraction Procedure for Antihistamine Drugs in Plasma (LLE Method)

Extraction Procedure

Plasma sample preparation begins with accurately transferring 2 mL of plasma into a clean 8 mL round-bottom threaded culture tube. An internal standard solution is added to each sample, followed by thorough vortex mixing to achieve uniform distribution. The sample is basified by adding 20 μL of 2 M sodium hydroxide, with additional mixing to ensure complete equilibration. Liquid–liquid extraction is carried out by adding 5 mL of a methyl tert-butyl ether and chloroform mixture (3:1, v/v). The tubes are securely capped and placed on a sample rotator to facilitate efficient partitioning of analytes into the organic phase. Phase separation is achieved by centrifugation, after which the organic layer is carefully transferred into a new clean tube without disturbing the aqueous phase. The organic extract is evaporated to dryness under a gentle stream of oxygen-free nitrogen at temperatures below 60 °C to prevent analyte degradation. The dried residue is reconstituted in 30 μL of a methanol-water mixture (1:1, v/v), vortex-mixed, and centrifuged for 5 minutes to remove particulates. The resulting clear supernatant is transferred into an autosampler vial for subsequent LC-MS analysis. Each analytical batch includes a biological blank, test samples, and a spiked control for quality assurance.

7.2.2 Non-steroidal Anti-inflammatory Drugs (NSAIDs)

NSAIDs are a diverse group of compounds characterized by their ability to inhibit COX enzymes, thereby reducing inflammation and pain [16, 17]. Structurally, NSAIDs vary from simple aromatic acids such as ibuprofen to more complex heterocyclic compounds like meloxicam. Their polarity ranges from non-polar (e.g., naproxen) to moderately polar (e.g., meloxicam), influencing both solubility and interaction with biological matrices.

In urine extraction, the polarity and pKa of NSAIDs play crucial roles. Non-polar NSAIDs often require organic solvents for efficient extraction, whereas polar compounds may benefit from SPE using anion exchange or mixed-mode sorbents. The presence of metabolites and the drug's ionization state at urinary pH can significantly influence recovery rates and detection sensitivity in LC–MS analyses. The molecular characteristics of the selected NSAIDs are summarized in Table 7.2.

Table 7.2 Molecular characteristics of selected NSAIDs

Drug name	Molecular formula	Molecular weight (g/mol)
Aspirin (Acetylsalicylic acid)	$C_9H_8O_4$	180.16
Ibuprofen	$C_{13}H_{18}O_2$	206.28
Naproxen	$C_{14}H_{14}O_3$	230.26
Ketoprofen	$C_{16}H_{14}O_3$	254.28
Diclofenac	$C_{14}H_{11}Cl_2NO_2$	296.15
Piroxicam	$C_{15}H_{13}N_3O_4S$	331.30
Meloxicam	$C_{14}H_{13}N_3O_4S_2$	351.40
Aceclofenac	$C_{16}H_{13}Cl_2NO_4$	354.19
Sulindac	$C_{20}H_{17}FO_3S$	356.41
Indomethacin	$C_{19}H_{16}ClNO_4$	357.79
Mefenamic acid	$C_{15}H_{15}NO_2$	241.28
Flurbiprofen	$C_{15}H_{13}FO_2$	244.26
Fenoprofen	$C_{15}H_{14}O_3$	242.27
Celecoxib	$C_{17}H_{14}F_3N_3O_2S$	381.37
Tolmetin	$C_15H_{15}NO_3$	257.29

Extraction Procedure for NSAIDs Analysis in Urine (SPE Method)

Urine samples are pre-treated to release the target analytes and to ensure compatibility with SPE [18, 19]. After pre-processing, the samples undergo SPE, which provides efficient isolation, concentration, and cleanup before LC–MS analysis.

Extraction Procedure

Sample Pre-treatment

Sample pre-treatment begins with accurately dispensing 6 mL of urine into a clean 50 mL glass beaker. Sodium hydroxide (100 μL, 10 M) is added to promote the hydrolysis of conjugated metabolites, followed by the addition of an appropriate internal standard solution to support accurate quantification and to compensate for potential matrix effects during LC–MS analysis. The mixture is incubated at 50 °C for 30 min to achieve complete base hydrolysis of the conjugated analytes. After incubation, the samples are allowed to cool to room temperature and are centrifuged to remove precipitated material. The clear supernatant is carefully transferred into a clean, labeled beaker, and the pH is adjusted to 6.8–7.0 using a suitable buffer to prepare the sample for solid-phase extraction.

Solid Phase Extraction

Solid-phase extraction is performed from 5 mL of the pre-traeted sample. The anion exchange SPE cartridge is conditioned sequentially with methanol and ammonium acetate buffer (pH 7.0) to activate and equilibrate the sorbent. The sample is further diluted with ammonium acetate buffer, mixed thoroughly, and passed through the cartridge to ensure efficient analyte retention. The cartridge is then washed with a methanol–ammonium acetate buffer mixture (1:9) and dried under a stream of oxygen-free nitrogen to remove residual moisture. Analytes are eluted using freshly prepared acetic acid in methanol (2%) to ensure complete recovery.

Post-Extraction Processing

Following extraction, the eluate evaporated to dryness under oxygen-free nitrogen at temperatures below 60 °C to prevent analyte degradation. The dried residue is reconstituted in a methanol-water mixture (1:1, v/v), vortex-mixed, and centrifuged to remove particulates. The resulting clear supernatant is transferred into an autosampler vial for subsequent LC-MS analysis.

Extraction Procedure for NSAIDs Analysis in Plasma (LLE Method)

Prior to extraction, blood samples are centrifuged to separate plasma, which is then accurately aliquoted to ensure consistent analyte recovery and reliable quantification [19]. The following protocol describes a simple liquid–liquid extraction (LLE) method for isolating NSAIDs from plasma while minimizing matrix effects.

Extraction Procedure

Plasma sample preparation involves transferring 2 mL of plasma into a clean, labeled 8 mL round-bottom threaded culture tube. An internal standard solution is added to each sample to ensure accurate quantification and to account for potential matrix effects. To extract the target analytes, 5 mL of methyl tert-butyl ether is added, and the tubes are securely capped. The samples are then mixed on a rotator at 40 rpm for 10 min to facilitate efficient partitioning of the analytes into the organic phase. Following centrifugation, the upper organic layer is carefully collected into a clean, labeled 8 mL tube, taking care to avoid disturbing the aqueous layer. The organic extract is evaporated to dryness under a stream of oxygen-free nitrogen at temperatures below 60 °C to maintain analyte stability. The resulting residue is reconstituted in 50 μL of a 1:1 methanol–water mixture, vortex-mixed, and centrifuged to remove any particulates. The clear supernatant is finally transferred into an autosampler vial for LC–MS analysis.

This extraction method offers a reliable and reproducible approach for isolating NSAIDs from urine and plasma prior to LC–MS analysis. Through the integration of pH adjustment, base hydrolysis, and carefully optimized SPE conditions, the protocol maximizes analyte recovery while minimizing matrix-related interferences. Its suitability for both urine and plasma allows for consistent and accurate detection and quantification of NSAIDs across various biological samples, supporting precise pharmacokinetic studies and forensic investigations.

7.2.3 Corticosteroids

Corticosteroids are steroidal hormones used as glucocorticoid or mineralocorticoid therapies, ranging from small polar molecules such as cortisol to bulky, lipophilic synthetic analogues like fluticasone [20–22]. Their physicochemical characteristics including molecular size, polarity, esterification, fluorination, and degree of protein binding significantly influence their extraction and detection in biological matrices. In urine, polar, acidic, or un-esterified corticosteroids are typically well recovered by reversed-phase SPE and show efficient ionization in LC MS analysis. Conversely, lipophilic, esterified, or prodrug forms often exhibit low parent concentrations in urine, necessitating optimized solvent systems, pH control, or targeted monitoring of metabolites for accurate quantification. In plasma samples, the high protein-binding affinity of corticosteroids poses additional challenges for extraction, requiring effective deproteinization and pH optimization to release the bound fraction. Careful selection of extraction solvents and SPE sorbents is therefore essential to achieve high recovery, minimize matrix effects, and ensure reliable LC–MS quantitation across both urine and plasma matrices. The molecular characteristics of the selected corticosteroid drugs are summarized in Table 7.3.

Table 7.3 Molecular characteristics of selected corticosteroids

Drug name	Molecular formula	Molecular weight (g·mol^{-1})
Cortisol (Hydrocortisone)	$C_{21}H_{30}O_5$	362.46
Prednisone	$C_{21}H_{26}O_5$	358.44
Prednisolone	$C_{21}H_{28}O_5$	360.45
Methylprednisolone	$C_{22}H_{30}O_5$	374.47
Dexamethasone	$C_{22}H_{29}FO_5$	392.46
Betamethasone	$C_{22}H_{29}FO_5$	392.46
Triamcinolone	$C_{21}H_{27}FO_6$	394.44
Triamcinolone acetonide	$C_{24}H_{31}FO_6$	434.50
Fludrocortisone	$C_{21}H_{29}FO_5$	380.46
Beclomethasone	$C_{22}H_{29}ClO_5$	408.92
Budesonide	$C_{25}H_{34}O_6$	430.54

Extraction Procedure for Corticosteroids Analysis in Urine (SPE Method)

Urine samples are processed using a SPE method to ensure efficient isolation and purification of analytes before LC–MS analysis [23, 24]. The procedure is optimized to improve recovery, minimize matrix interference, and maintain analytical reproducibility.

Extraction Procedure

Sample Pre-treatment

The extraction procedure begins by accurately dispensing 7 mL of urine into a clean beaker and adjusting the pH to 6.8–7.0 using an appropriate buffer to optimize extraction efficiency. Enzymatic deconjugation is performed by adding β-glucuronidase from *Escherichia coli* (2000 units/mL) and incubating overnight at 37 °C. Following deconjugation, the sample is centrifuged to remove particulate matter and obtain a clear supernatant. 5 mL of this supernatant are transferred to a clean test tube, and suitable internal standard solution is added, followed by thorough vortexing to ensure complete homogenization.

Solid Phase Extraction

A non-polar C18 SPE cartridge is conditioned using 2 mL of methanol followed by 2 mL of deionized water to activate and equilibrate the sorbent. The prepared sample is then passed through the cartridge at a flow rate not exceeding 1 mL/min, allowing efficient adsorption of analytes onto the sorbent. Retained analytes are eluted with 5 mL of a dichloromethane: ethyl acetate mixture (4:1, v/v) at a controlled flow rate, and the eluate is washed with 1 mL of 0.2 M NaOH to remove residual interferences.

Post-Extraction Processing

The organic layer is carefully collected into a clean, dry 8 mL round-bottom threaded tube, avoiding any aqueous phase carryover. The solvent is evaporated to dryness under a gentle stream of oxygen-free nitrogen at 60 °C, and the residue is reconstituted in 50 μL of a 1:1 methanol–water mixture to prepare the extract for LC–MS analysis.

This protocol produces a clean and concentrated analyte fraction suitable for high-sensitivity LC–MS detection. The combination of C18-based SPE and optimized elution conditions ensures reproducible recovery with minimal matrix interference, enhancing the reliability of drug quantification in urine samples.

Extraction Procedure for Corticosteroids Analysis in Plasma (LLE Method)

Plasma samples are processed to efficiently isolate target analytes while minimizing matrix effects before LC–MS analysis [23, 24]. The method combines liquid–liquid extraction with careful solvent handling to achieve high recovery and reproducible results.

Extraction Procedure

The plasma extraction procedure begins by transferring 1 mL of plasma into a clearly labeled 8 mL round-bottom threaded culture tube. A suitable internal standard is added to each tube, and the sample is vortexed thoroughly to ensure complete homogenization. To release corticosteroids from plasma proteins, 1 mL of 4% phosphoric acid is added, facilitating protein dissociation. Liquid–liquid extraction is then performed by adding 4 mL of methyl tert-butyl ether (MTBE), allowing the analytes to partition into the organic phase. The tubes are placed on a sample rotator at 40 rpm for 10 min to achieve thorough mixing, followed by centrifugation to obtain clear phase separation. The upper organic layer is carefully collected into a clean, labeled 8 mL tube, avoiding any carryover of the aqueous phase. The organic extract is evaporated to dryness under a gentle stream of oxygen-free nitrogen at 60 °C, and the residue is reconstituted in 50 μL of a 1:1 methanol–water mixture, preparing the sample for subsequent LC–MS analysis.

7.2.4 Anabolic Steroids

Anabolic–androgenic steroids (AAS) are steroid hormones, either naturally occurring or synthetic derivatives of testosterone, that primarily promote muscle growth, protein synthesis, and tissue development [25, 26]. Their structural and physicochemical properties, including polarity, lipophilicity, molecular size, and protein-binding affinity, strongly influence their extraction and analysis from biological matrices. Polar AAS are more water-soluble, allowing relatively straightforward extraction from urine using techniques such as SPE or LLE. In contrast, lipophilic steroids, which often bind to plasma proteins, require more elaborate solvent-based extraction methods to achieve efficient recovery. Additionally, anabolic steroids undergo extensive metabolic transformations, producing phase I and phase II metabolites that must be considered for comprehensive detection. Understanding these characteristics is critical for optimizing analytical workflows, such as LC–MS/MS, to ensure accurate quantification, reduce matrix interference, and reliably detect both parent compounds and metabolites. Molecular properties of selected anabolic steroids are summarized in Table 7.4.

Optimized Urine Extraction and Hydrolysis Protocol for Anabolic Steroid Analysis via LC–MS

Urine samples are processed using a combination of SPE and LLE to efficiently isolate steroid analytes while minimizing matrix interferences. The workflow maximizes recovery, preserves analyte integrity, and ensures compatibility with high-sensitivity LC–MS detection.

Extraction Procedure

Extraction—Step 1

The extraction procedure begins with centrifuging the urine sample at 5500 rpm for 20 min, after which 3 mL of the clear supernatant is transferred into a clean test tube. A suitable internal standard is added, and the sample is vortexed thoroughly to ensure uniform distribution. The sample is then diluted with 4 mL of deionized water to optimize loading onto the SPE cartridge. Biological blanks, calibration standards, spiked controls, and samples are loaded onto the SPE cartridge. The polymeric, non-polar/reversed-phase SPE cartridge is conditioned sequentially with 3 mL of methanol followed by 3 mL of water to activate and equilibrate the sorbent. The diluted sample is passed through the cartridge, which is subsequently rinsed with 4 mL of water followed by 3 mL of hexane and then dried under oxygen-free nitrogen

Table 7.4 Molecular characteristics of selected anabolic steroids

Steroid	Molecular formula	Molecular weight (g/mol)
Testosterone	$C_{19}H_{28}O_2$	288.43
Nandrolone	$C_{18}H_{26}O_2$	274.40
Boldenone	$C_{19}H_{26}O_2$	286.42
Stanozolol	$C_{21}H_{32}N_2O$	328.50
Methenolone	$C_{20}H_{30}O_2$	302.46
Oxandrolone	$C_{19}H_{30}O_3$	306.45
Oxymetholone	$C_{21}H_{32}O_3$	332.48
Trenbolone	$C_{18}H_{22}O_2$	277.37
Fluoxymesterone	$C_{20}H_{29}FO_3$	336.45
Bolasterone	$C_{21}H_{32}O_2$	316.49
5α-Dihydrotestosterone	$C_{19}H_{30}O_2$	290.45

for 2 min. The analytes are eluted with 4 mL of methanol into a clean, labeled 8 mL round-bottom threaded tube for further processing.

Hydrolysis Step

For hydrolysis, the eluate is transferred to an 8 mL round-bottom tube and evaporated to dryness at 60 °C under oxygen-free nitrogen. Methanolic-HCl (0.5 mL, 1.0 M) is added, the tube is vortexed, capped, and incubated at 60 °C for 15 min. Residual acid is neutralized with 2 mL of a caustic wash solution prepared by dissolving 20 g NaOH and 14.6 g NaCl in 500 mL of distilled water.

Extraction—Step 2

Liquid–liquid extraction is then performed by adding 4 mL of di-isopropyl ether to the hydrolyzed sample, which is capped and rotated at 40 rpm for 10 min to ensure thorough mixing. The mixture is centrifuged, and the organic layer is carefully collected into a clean, labeled 8 mL round-bottom tube. The collected organic phase is evaporated to dryness under nitrogen at 60 °C and reconstituted in a suitable solvent for subsequent LC–MS analysis.

SPE of Anabolic Steroid from Plasma Samples for LC–MS Analysis

The following SPE protocol provides a reliable and reproducible method to obtain clean extracts from plasma samples with minimal matrix effects, suitable for both qualitative and quantitative determination of anabolic steroid.

Extraction Procedure

Sample Pre-treatment

Sample pre-treatment begins by dispensing 0.5 mL of each biological blank, test sample, and spiked control into labeled 2 mL microcentrifuge tubes. Plasma proteins are precipitated by adding 1 mL of ice-cold acetonitrile, followed by thorough vortexing and centrifugation. The clear supernatant is carefully transferred into clean, labeled 8 mL round-bottom threaded culture tubes, and 5 mL of deionized water is added to each tube with subsequent vortexing. A suitable internal standard is added to each sample to ensure accurate quantification, and the mixture is gently vortexed to achieve homogeneity. The biological blanks, test samples, and spiked controls are then taken for solid-phase extraction (SPE).

Solid Phase Extraction

The SPE process begins with conditioning the polymeric non-polar/reversed-phase cartridge by sequentially priming it with 3 mL of methanol followed by 3 mL of water to activate and equilibrate the sorbent. The pre-treated sample is loaded onto the pre-conditioned cartridge. The cartridge is rinsed sequentially with 4 mL of water and 3 mL of hexane, then dried under oxygen-free nitrogen for 2 min. Analytes are eluted with 4 mL of methanol into a clean, labeled 8 mL round-bottom tube.

Post-Extraction Processing

The organic layer is evaporated under oxygen-free nitrogen at temperatures below 60 °C to preserve analyte integrity. The resulting residue is reconstituted in 50 μL of a 1:1 methanol–water mixture, vortexed thoroughly, and centrifuged to remove any particulates. The clear supernatant is then transferred into an autosampler vial for LC–MS analysis.

7.2.5 Peptide Drugs

Peptide drugs are short amino acid chains that mimic or modulate natural physiological hormones and signaling molecules. Their large molecular size, polarity, and susceptibility to enzymatic degradation make extraction from biological matrices challenging [27, 28]. In urine, peptides often appear as small fragments or metabolites, requiring specialized solid-phase or immunoaffinity extraction for reliable LC–MS detection. The molecular characteristics of the selected peptides are summarized in Table 7.5.

SPE Procedure for Peptide Drug Analysis in Urine and Plasma

Efficient extraction of peptide-based drugs from biological matrices such as urine or plasma is crucial for achieving high analytical sensitivity and reproducibility.

Table 7.5 Molecular characteristics of selected peptides

Drug name	Molecular formula	Molecular weight	Precursor Ion (m/z)	Product Ion 1 (m/z)	Product Ion 2 (m/z)	Product Ion 3 (m/z)
Bremelanotide	$C_{50}H_{68}N_{14}O_{10}$	1024.5243	513.40	413.40	435.90	497.47
Buserelin	$C_{60}H_{86}N_{16}O_{13}$	1238.6555	620.40	299.20	592.25	749.40
Capromorelin	$C_{28}H_{35}N_5O_4$	505.2689	506.20	235.12	244.11	263.12
Dermorphin	$C_{40}H_{50}N_8O_{10}$	802.8700	803.37	602.20	202.10	574.30
GHRP-2	$C_{45}H_{55}N_9O_6$	817.4270	409.70	269.10	479.20	550.30
GHRP-3	$C_{32}H_{50}N_{10}O_5$	654.3960	328.30	244.10	272.20	384.30
GHRP-6	$C_{46}H_{56}N_{12}O_6$	872.4440	437.20	324.10	581.30	728.30
TB-500	$C_{38}H_{68}N_{10}O_{14}$	888.4911	445.40	129.10	606.20	734.40
GHRP-4	$C_{34}H_{38}N_7O_4$	608.2974	609.40	444.20	592.30	352.20
TAPS	$C_{27}H_{37}N_7O_6$	555.2800	556.40	303.10	321.00	485.40
Deslorelin	$C_{64}H_{83}N_{17}O_{12}$	1281.6402	641.80	299.20	598.40	848.50
GHRP-1	$C_{51}H_{62}N_{12}O_7$	954.4859	319.20	225.10	293.20	479.30
Histrelin	$C_{66}H_{86}N_{18}O_{12}$	1322.6656	441.90	249.00	412.30	538.30

Extraction Procedure

Sample Pre-treatment

Sample pre-treatment begins by dispensing 2.5–3 mL of the biological sample (urine or plasma) into a clean test tube and adjusting the pH to 6.8–7.0 to optimize peptide retention. The sample is centrifuged at 5500 rpm for 15 min to remove particulates, and 2 mL of the clear supernatant is transferred into a clean, labeled test tube. A suitable internal standard solution is added to each sample and vortexed thoroughly to ensure homogeneity. The processed samples are then taken for SPE.

Solid Phase Extraction

For solid-phase extraction, mixed-mode weak anion-exchange polymeric SPE cartridges are conditioned sequentially with 2 mL of methanol followed by 2 mL of 1.25% aqueous ammonia. 3 mL of 1.25% aqueous ammonia are added to the pre-treated sample and mixed thoroughly before passing it through the cartridge at a controlled flow rate. The cartridge is rinsed with 2.5 mL of 1.25% aqueous ammonia followed by 2.5 mL of a 1:1 methanol–water mixture to remove weakly bound impurities and then dried under a stream of nitrogen for 1 min. Retained peptides are eluted slowly with 2 mL of 5% formic acid in methanol at 0.02 mL/sec into clean, labeled 8 mL round-bottom threaded tubes.

Post-Extraction Processing

The eluate is evaporated to dryness under a gentle stream of oxygen-free nitrogen below 60 °C to prevent peptide degradation. The residue is reconstituted in 50 μL of 2% acetic acid in water to enhance peptide ionization and LC–MS compatibility, vortex-mixed, and centrifuged at 2000 rpm for 5 min to remove any residual partic ulates. The clear supernatant is transferred into an autosampler vial for LC–MS analysis. Each analytical batch includes a biological blank, spiked control, and test samples to ensure method accuracy and quality assurance.

7.2.6 Bisphosphonate and Myo-Inositol Trispyrophosphate (ITPP) Drugs

Bisphosphonate and ITPP drugs are highly polar and strongly hydrophilic compounds with multiple phosphate or phosphonate groups, resulting in poor retention on non-polar extraction media [29–31]. In biological matrices, they exist predominantly in ionized forms, making conventional LLE inefficient. SPE using anion-exchange or mixed-mode sorbents is preferred to enhance selectivity and reproducibility.

The selected bisphosphonate compounds along with respective molecular formulas, molecular weights, charge states, and characteristic fragment ions detected under LC–MS conditions, are summarized in Table 7.6. These parameters provide essential information for the reliable identification and quantification of bisphosphonate drugs in biological matrices such as urine and plasma.

SPE for Bisphosphonate Drug Analysis (Urine/Plasma)

The main challenge in the analysis of bisphosphonates is their poor LC–MS compatibility. By integrating mixed-mode solid-phase extraction with a derivatization step, the procedure enhances chromatographic retention and improves detection sensitivity in mass spectrometric analysis. This optimized approach provides consistent, accurate, and reproducible results, making it highly suitable for use in forensic investigations and anti-doping testing.

Extraction Procedure

Table 7.6 Important bisphosphonate drugs with their molecular characteristics and LC–MS fragmentation data

Drug name	Derivative	Molecular formula	Molecular weight	Precursor Ion (m/z)	Fragment Ions (m/z)
Etidronic acid	Etidronic acid-4Methyl	$C_6H_{16}O_7P_2$	262.0366	263.0444	137.0361, 127.0155, 165.0422
Ibandronic acid	Ibandronic acid-4Methyl	$C_{13}H_{31}NO_7P_2$	375.1570	376.1649	344.1375, 275.0437, 250.1560
Tiludronic acid	Tiludronic acid-4Methyl	$C_{11}H_{17}ClO_6P_2S$	373.9904	374.9982	342.9712, 156.9873, 281.1439
Pamidronic acid	Pamidronic acid-7Methyl	$C_{10}H_{25}NO_7P_2$	333.1101	334.1179	275.0433, 208.1092, 149.0359
Neridronic acid	Neridronic acid-7Methyl	$C_{13}H_{31}NO_7P_2$	375.1570	376.1649	317.0903, 250.1553, 191.0830
Alendronic acid	Alendronic acid-7Methyl	$C_{11}H_{27}NO_7P_2$	347.1257	348.1336	348.1328, 289.0596, 163.0517
Zoledronic acid	Zoledronic acid-4Methyl	$C_9H_{18}N_2O_7P_2$	328.0584	329.0662	261.0284, 203.0578, 288.0447
Clodronic acid	Clodronic acid-4Methyl	$C_5H_{12}Cl_2O_6P_2$	299.9481	300.9559	268.9290, 236.9838, 127.0155
Risedronic acid	Risedronic acid-4Methyl	$C_{11}H_{19}NO_7P_2$	339.0631	340.0710	323.0315, 214.0625, 182.0363

Extraction Steps

Extraction begins by dispensing 2 mL of centrifuged urine or plasma into a clean, labeled test tube. A suitable internal standard is added, followed by 2 mL of 2% formic acid, and the sample is vortexed thoroughly to ensure complete mixing. A hydrophilic-lipophilic balanced (HLB) polymeric sorbent cartridge is conditioned sequentially with 2 mL of methanol and 2 mL of deionized water. The prepared sample is passed slowly through the HLB cartridge, and the filtrate is collected in a clean test tube. The collected filtrate is then passed through a pre-conditioned polymeric reversed-phase, weak anion-exchange mixed-mode sorbent cartridge, previously conditioned with 2 mL methanol and 2 mL of 2% formic acid. The cartridge is washed sequentially with 3 mL of 2% formic acid and 3 mL of methanol to remove impurities, then dried completely under gentle vacuum or oxygen-free nitrogen. Analytes are eluted with 4 mL of 3.5% ammonia in methanol into a clean, dry 8 mL round-bottom threaded tube, and the eluate is evaporated to dryness at 45 °C under oxygen-free nitrogen to obtain the residue for derivatization.

Derivatization

For derivatization, either the TMS-diazomethane or trimethylorthoacetate method can be applied. In the TMS-diazomethane approach, 50 μL of water is added and vortexed, followed by 100 μL each of methanol and trimethylsilyldiazomethane. The mixture is capped, mixed thoroughly, and incubated at room temperature for 1 h to complete methylation. The sample is then evaporated to dryness under oxygen-free nitrogen at 45 °C and reconstituted in 100 μL of a 1:1 methanol–water mixture before transfer to autosampler vials for LC–MS analysis.

In the trimethylorthoacetate approach, 50 μL of glacial acetic acid and 400 μL of trimethylorthoacetate are added to the residue, the tube is capped, vortexed, and heated at 100 °C for 60 min to ensure complete derivatization. The contents are evaporated to dryness at 45 °C under oxygen-free nitrogen, reconstituted in 100 μL of a 1:1 methanol–water mixture, and transferred to autosampler vials for LC–MS analysis.

This dual-cartridge extraction and derivatization protocol provides an efficient, selective, and reproducible approach for detecting bisphosphonates in complex biological matrices. The optimized method ensures high analytical sensitivity and specificity, facilitating accurate detection and quantification in forensic and anti-doping investigations.

Extraction Procedure for ITPP in Urine and Plasma

The molecular properties and LC–MS fragmentation behavior of ITPP are summarized in Table 7.7, which outlines key parameters including molecular formula, molecular weight, and major fragment ions.

Table 7.7 Molecular characteristics and LC–MS fragmentation data of ITPP

Drug name	Mol. formula	Mol. Wt	Precursor Ion	Product Ions
ITPP	$C_6H_{12}O_{21}P_6$	605.9	302	525, 445, 79
ITPP-d_6 (IS)	$C_6D_6H_6O_{21}P_6$	611.9	305	531, 451, 239

This extraction protocol was specifically designed to isolate and concentrate ITPP from biological matrices such as urine and plasma. The method employs a two-step SPE approach combining HLB and polymeric reversed-phase, weak anion-exchange mixed-mode sorbent cartridges, ensuring optimal recovery of this highly polar compound while minimizing matrix interferences prior to LC–MS analysis.

Extraction Procedure

The extraction procedure begins by transferring 2 mL of the centrifuged biological sample into a clean, labeled test tube. A suitable internal standard is added, followed by the addition of 2 mL of 2% formic acid, and the mixture is vortexed thoroughly to ensure effective acidification and homogeneity. A hydrophilic–lipophilic balanced (HLB) polymeric sorbent cartridge is conditioned sequentially with 2 mL of methanol and 2 mL of deionized water. The pre-treated sample is passed slowly through the HLB cartridge, and the resulting filtrate is collected in a clean test tube. A polymeric reversed-phase, weak anion-exchange mixed-mode sorbent cartridge is preconditioned with 2 mL of methanol followed by 2 mL of 2% formic acid, after which the filtrate is loaded and allowed to pass slowly through the cartridge. The cartridge is washed with 3 mL of 2% formic acid followed by 3 mL of methanol to remove weakly retained impurities and then dried completely under a gentle stream of nitrogen gas. The retained analytes are eluted with 4 mL of 3.5% ammonia in methanol into a clean, dry 8 mL round-bottom threaded culture tube. The eluate is evaporated to dryness under oxygen-free nitrogen at 45 °C, and the residue is reconstituted in 50 μL of a solvent mixture consisting of 25 mM ammonium acetate, methanol, and acetonitrile (10:45:45, v/v/v). The solution is gently vortexed, transferred into an autosampler vial, and subjected to LC–MS analysis.

This optimized SPE-based protocol ensures efficient extraction of ITPP from complex biological fluids with minimal degradation or ion suppression. The combined HLB-Polymeric reversed-phase, weak anion-exchange mixed-mode sorbent cartridges approach provides high selectivity, making it particularly suitable for trace-level quantification.

7.2.7 *Quaternary Ammonium Drugs*

Quaternary ammonium compounds carry a permanent positive charge on their nitrogen atom, which makes them highly polar and readily soluble in water [32]. Because of this strong ionic character, they show low lipid solubility, poor membrane permeability, and very limited ability to cross into the central nervous system. These properties also make their extraction from urine and plasma difficult, as their high polarity and constant ionization reduce their interaction with typical SPE or LLE solvents. Because conventional extraction methods often yield low recoveries, these drugs are most reliably processed using a dilute-and-inject strategy. This approach minimizes sample loss, simplifies preparation, and supports accurate quantification of these highly water-soluble ionic substances. The molecular characteristics and LC–MS fragmentation data of key quaternary ammonium drugs are summarized in Table 7.8.

Extraction Procedure for Quaternary Ammonium Drugs in Urine and Plasma

Extraction Procedure

The extraction procedure begins by dispensing 20 μL of the biological sample (urine or plasma) into a 1.5 mL microcentrifuge tube. A suitable internal standard solution (20 μL) is added, followed by the addition of 70 μL of acetonitrile and 90 μL of 10 mM ammonium formate solution. The mixture is vortexed thoroughly to ensure complete homogenization and protein precipitation. The sample is then centrifuged to allow particulate matter to settle. Subsequently, 100 μL of the clear supernatant is transferred into pre-labeled autosampler vials for LC–MS analysis under optimized chromatographic conditions.

Each analytical batch includes a biological blank, test samples, a spiked control, to ensure analytical reliability and method accuracy.

This streamlined extraction and direct injection protocol provides a rapid, reproducible, and solvent-efficient approach suitable for high-throughput analysis of quaternary ammonium drugs in biological samples.

Table 7.8 Selected quaternary ammonium drugs: Molecular characteristics and LC–MS fragmentation data

Drug name	Mol. formula	Mol. Wt	Precursor Ion	Fragment Ions (1)	Fragment Ions (2)	Fragment Ions (3)
Ipratropium	$C_{20}H_{30}NO_3$	332.2220	332.2220	166.1691	124.1123	290.1750
Mebezonium	$C_{19}H_{40}N_2$	296.3186	148.1590	236.2375	60.0816	194.1898
Meldonium	$C_6H_{14}N_2O_2$	146.1050	147.1128	58.0662	59.0739	132.0895
N-Butylscopolammonium	$C_{21}H_{30}NO_4$	360.2169	360.2169	194.1542	138.0916	121.0652
Pancuronium	$C_{35}H_{60}N_2O_4$	572.4548	286.2276	236.6750	206.6644	430.3320
Tiotropium	$C_{19}H_{22}NO_4S_2$	392.0985	392.0985	152.1071	170.0176	122.0967
Clidinium	$C_{22}H_{26}NO_3$	352.1907	352.1907	142.1228	124.1244	96.0813
Rocuronium	$C_{32}H_{53}N_2O_4$	529.4000	529.3999	487.3534	112.1125	358.2738
Oxyphenonium	$C_{21}H_{34}NO_3$	348.2533	348.2533	132.1384	149.0958	261.1483
Mivacurium	$C_{58}H_{80}N_2O_{14}$	1028.5604	514.2799	342.1993	189.0913	428.2438
Glycopyrrolate	$C_{19}H_{28}NO_3$	318.2064	318.2064	116.1076	98.0970	58.0661

7.3 Standardized Protocols for Extraction and Isolation of Drugs from Hair Samples for LC–MS Analysis

Hair testing is widely used as a non-invasive approach for identifying drug use, environmental exposure, and long-term toxicological patterns [33, 34]. Unlike blood or urine, which reflect only recent intake, hair provides an extended detection window because substances become incorporated into the hair shaft as it grows. This feature makes hair a valuable matrix in forensic, clinical, and anti-doping settings, where evidence of repeated or chronic exposure is required [35]. However, its complex keratin structure and the strong binding of many compounds mean that thorough decontamination and careful extraction procedures are necessary to ensure accurate and reliable analytical results.

Extraction Procedure

Segmentation

Hair sample preparation begins with segmentation of the strands into proximal and distal portions based on sample length, with each segment carefully labeled and processed separately.

Decontamination Procedure

Approximately 500 mg of each hair segment is weighed into a clean beaker and subjected to a sequential decontamination procedure. The hair is washed with 5 mL of 0.1% sodium dodecyl sulfate (SDS), sonicated for 5 min, and the solution is decanted; this step is repeated three times to remove surface contaminants. The sample is subsequently washed three times with 5 mL of deionized water, followed by three washes with 5 mL of methanol–water (1:4, v/v), with sonication and careful decanting at each step. The decontaminated hair is air-dried overnight at room temperature prior to further processing.

Pulverization and Extraction

For pulverization and extraction, the dried hair strands are cut into small segments (2–3 mm) and transferred to a 50 mL beaker. An aliquot of 100 mg is placed into 7 mL bead ruptor tubes, and a suitable internal standard is added. Mechanical pulverization is performed using a bead mill homogenizer operated at 5 m/s for 50 s with 15-s rest intervals over six cycles. Methanol (1.5 mL) is then added, followed by additional homogenization at 4 m/s for 50 s with 15-s rest intervals over two cycles. The samples are centrifuged, and 1.2 mL of the supernatant is transferred into 1.5 mL microcentrifuge tubes. These tubes are centrifuged at 15,000 rpm for 10 min. The extraction is repeated, and the supernatants are combined in a clean, labeled 8 mL round-bottom threaded culture tube. 50 μL of dimethyl sulfoxide (DMSO) is added, and the extract is evaporated under a gentle stream of nitrogen at 60 °C to a final volume of approximately 50 μL. Subsequently, 3 mL of 0.1 M phosphate buffer is added, the mixture is vortexed, and the sample is prepared for solid-phase extraction (SPE).

Solid Phase Extraction

For SPE cleanup, polymeric non-polar/reversed-phase cartridges are conditioned sequentially with 2 mL of methanol, 2 mL of water, and 2 mL of 0.1 M phosphate buffer. The prepared sample is loaded onto the cartridge and passed through slowly to ensure efficient analyte retention. The cartridge is washed with 2 mL of water followed by 2 mL of hexane and dried under vacuum for 1 min. Analytes are eluted twice with 3 mL of methanol to ensure complete recovery.

Post-Extraction Processing

The combined eluates are evaporated to dryness under oxygen-free nitrogen at 60 °C and reconstituted in 50 μL of methanol–water (1:4, v/v). After vortexing and centrifugation at 2000 rpm for 5 min, the clear supernatant is transferred into autosampler vials for LC–MS analysis under validated chromatographic conditions.

This comprehensive extraction and purification approach effectively removes external contamination while ensuring efficient analyte recovery from the complex hair matrix. The integration of mechanical homogenization and SPE cleanup enhances analytical sensitivity, reproducibility, and suitability for forensic, clinical, and anti-doping applications.

7.4 Sample Preparation and Extraction Techniques for Drug Detection by GC–MS

GC–MS remains a preferred analytical tool for detecting specific drug classes that show poor ionization efficiency in LC–MS. Non-polar compounds, including anabolic steroids (e.g., Estranediol and androstanediol), lack readily ionizable functional groups, resulting in weak LC–MS signals [34, 36].

Through derivatization commonly using reagents like N-methyl-N (trimethylsilyl) trifluoroacetamide (MSTFA) or N, O-Bis(trimethylsilyl)trifluoroacetamide (BSTFA) polar groups are transformed into volatile, thermally stable silylated derivatives, improving separation and detection in GC–MS. This approach enhances analytical sensitivity, selectivity, and structural confirmation of the target molecules. Moreover, since biological matrices such as urine and plasma contain complex interferences, efficient extraction before GC–MS analysis is essential to isolate drugs and achieve accurate quantification, making this technique invaluable in forensic, clinical, and pharmacokinetic investigations. The principal anabolic drugs identified through GC–MS analysis are listed in Table 7.9, representing compounds that exhibit suitable volatility and thermal stability upon derivatization for effective chromatographic separation and mass spectrometric detection.

Table 7.9 List of selected anabolic drugs detectable by GC–MS

S. no	Drug name
1	17 Epimethandienone
2	Boldione
3	5α-Estrane Diol
4	Androstane
5	Androstene
6	
7	Bolasterone
8	Boldenone
9	Clostebol
10	Dehydroepiandrosterone
11	Epitestosterone
12	Fluoxymesterone
13	Mestanalone
14	Mestranol
15	Methandriol
16	Methyl Testosterone
17	Nandrolone
18	Norbolethone
19	Norethandralone
20	Normethandrolone
21	Oxandrolone
22	Stanolone
23	Testosterone
24	Boldione
25	Trenbolone
26	Methylandrostane Diol
27	19-Nor Androsterone
28	Androst 4 Ene 3,17 Dione
29	17 Epimethandienone
30	Methandrostenolone

7.4.1 *Sample Preparation of Anabolic Steroids and Their Extraction from Urine for GC–MS Detection*

Solid phase extraction offers an efficient and reproducible approach for isolating drug analytes from complex biological matrices such as urine/plasma. This protocol describes an optimized SPE method designed to prepare urine samples for subsequent GC–MS analysis, enabling accurate quantification of target compounds.

Extraction Procedure

Sample Pre-treatment

Each analytical batch comprises a biological blank, test samples, and a spiked control. The extraction procedure begins by centrifuging urine samples to remove particulate matter. An aliquot of 3 mL of the clarified urine is transferred into a clean test tube, a suitable internal standard is added, and the sample is vortex-mixed to ensure homogeneity. Deionized water (4 mL) is then added, followed by additional vortexing for complete mixing. The biological blank, calibration standards, spiked control, and test samples are loaded onto an SPE cartridge.

Solid Phase Extraction

Polymeric non-polar/reversed-phase SPE cartridges are conditioned sequentially with 3 mL of methanol and 3 mL of water. The pre-treated sample is loaded onto the cartridge and allowed to pass slowly to ensure complete analyte retention. The cartridges are subsequently washed with 4 mL of water followed by 3 mL of hexane to remove matrix interferences and are dried under oxygen-free nitrogen for 2 min. Retained analytes are eluted with 4 mL of methanol into clean, labeled 8 mL round-bottom threaded culture tubes.

Hydrolysis

For hydrolysis, the eluates are evaporated to dryness under oxygen-free nitrogen at 60 °C. Methanolic hydrochloric acid (0.5 mL, 1 M) is added, and the samples are capped, vortexed, and heated at 60 °C for 15 min in a fume hood, ensuring that the incubation time does not exceed 15 min to prevent steroid degradation. Following hydrolysis, 2 mL of a caustic wash solution prepared is added to neutralize residual acid and prepare the samples for further extraction.

Liquid Phase Extraction

A second extraction is performed by adding 4 mL of di-isopropyl ether and rotating the tubes at 40 rpm for 10 min to facilitate thorough mixing. After centrifugation to achieve clear phase separation, the organic layer is carefully transferred into clean, labeled 8 mL round-bottom tubes and evaporated to dryness at 60 °C under oxygen-free nitrogen to obtain residues suitable for derivatization.

Derivatization

For PFPA derivatization, 70 μL of anhydrous acetonitrile, pre-dried using molecular sieves, and 30 μL of pentafluoropropionic anhydride are added to the residue. The mixture is incubated at 60 °C for 15 min, evaporated to dryness at 80 °C under nitrogen, and reconstituted in 50 μL of ethyl acetate before transfer to autosampler vials for GC–MS analysis. Alternatively, enol-TMS derivatization is performed by adding 50 μL of Enol-TMS reagent to the dried extract and incubating at 60 °C for 30 min, after which the sample is transferred directly into autosampler vials for GC–MS analysis.

This SPE procedure provides a robust, reproducible, and high-throughput approach for preparing urine samples prior to GC–MS analysis. The integration of hydrolysis, extraction, and derivatization steps ensures efficient removal of interferences and conversion of non-volatile analytes, into stable derivatives suitable for precise detection. This streamlined method significantly enhances analytical reliability for forensic, clinical, and toxicological studies.

7.4.2 Sample Preparation of Amphetamine and Their Extraction from Urine for GC–MS Detection

This protocol outlines the preparation, enzymatic hydrolysis, and solid-phase extraction of urine samples for the detection of amphetamines using GC–MS. Each batch includes internal standards, spiked controls, and biological blanks, with derivatization performed to enhance detection and ensure accurate quantification.

Extraction Procedure

Sample Preparation

Sample preparation begins by dispensing 5 mL of urine into a 50 mL beaker and adjusting the pH to 5.0. A suitable internal standard solution is added, followed by the addition of 30 μL of trypsin and β-glucuronidase from *Helix pomatia* (equivalent to 10,000 units per sample). The mixture is thoroughly mixed and incubated either overnight at 37 °C or for 2 h at 56 °C to achieve enzymatic hydrolysis. After incubation, the pH is adjusted to 6.0–6.5 using dilute ammonium hydroxide, and the sample is centrifuged to remove particulate matter.

Solid Phase Extraction

Solid-phase extraction is performed using a mixed-mode strong cation-exchange cartridge, which is conditioned sequentially with 2 mL of methanol followed by 2 mL of 0.1 M phosphate buffer. The sample is diluted with 2 mL of phosphate buffer, mixed, and is passed through the cartridge. The cartridge is washed with 1 mL of 0.1 M phosphate buffer and 2 mL of 1 M acetic acid, then dried under a stream of nitrogen for 1 min. A further rinse with 1 mL of methanol is performed, followed by drying under nitrogen for an additional 2 min. Analytes are eluted using 5 mL of freshly prepared ethyl acetate–ammonia–isopropanol (96:2:2, v/v/v).

Derivatization and GC–MS Preparation

The eluate is transferred into an 8 mL round-bottom threaded culture tube and evaporated to dryness under oxygen-free nitrogen at temperatures below 60 °C. Derivatization is carried out by adding 30 μL of BSTFA containing 10% TMCS and incubating at 60 °C for 20 min with vortex mixing. The resulting trimethylsilyl derivatives are transferred into vials for GC–MS analysis.

7.4.3 Sample Preparation of Caffeine and Theophylline and Their Extraction from Urine for GC–MS Detection

This protocol describes the preparation, base hydrolysis, and solid-phase extraction of urine samples for the detection of caffeine and theophylline using GC–MS. Each batch includes biological blanks, samples, and spiked controls, with derivatization to enhance detection and ensure accurate quantification.

Extraction Procedure

Sample Preparation and Hydrolysis

Sample preparation begins by dispensing 7 mL of urine into a 50 mL beaker and adding 100 μL of 10 M sodium hydroxide. Base hydrolysis is carried out at 50 ± 5 °C for 30 min to cleave conjugated analytes. The samples are then allowed to cool, and the pH is adjusted to 7.0. Following pH adjustment, the samples are centrifuged to remove any solid residues.

Solid Phase Extraction

An aliquot of 5 mL of the pre-treated sample is transferred to clean test tube, and a suitable internal standard is added. Solid-phase extraction is performed using a mixed-mode strong cation-exchange cartridge, which is conditioned sequentially with 3 mL of methanol followed by 3 mL of 0.05 M ammonium acetate buffer (pH 7.0). The sample is diluted with 2 mL of ammonium acetate buffer, mixed thoroughly, and passed through the cartridge. The cartridge is washed with 1 mL of methanol-ammonium acetate buffer (1:9, v/v) to remove matrix interferences and then dried under a stream of nitrogen for 2 min. Analytes are eluted using 5 mL of freshly prepared 2% acetic acid in methanol.

Derivatization and GC–MS Preparation

For derivatization and GC–MS analysis, the eluate is transferred into an 8 mL round-bottom threaded culture tube and evaporated to dryness under nitrogen at temperatures below 60 °C. The residue is treated with 200 μL of acetone, a small amount of potassium carbonate, and 30 μL of iodomethane (or iodoethane with diisopropylamine for theophylline confirmation). The mixture is vortex-mixed and incubated at 60 ± 5 °C for 30 minutes. After cooling, the samples are evaporated to dryness under nitrogen, reconstituted with 50 μL of toluene, vortex-mixed, and centrifuged. The clear solution is transferred into autosampler vials for GC-MS analysis.

7.5 Method Validation in Forensic and Anti-doping Testing

In forensic toxicology and anti-doping analysis, method validation is essential to ensure that analytical procedures are reliable, reproducible, accurate, and fit for the intended purpose. Since these domains frequently involve legal, regulatory, or high-stake decisions (e.g. suspension of athletes, legal consequences, public safety), validation ensures defensible results. Validation covers a suite of parameters selectivity, linearity, limits of detection and quantitation (LOD, LOQ), recovery, matrix effects, accuracy, precision, among others [37–39]. International and sector-specific guidelines like Scientific Working Group for Forensic Toxicology (SWGTOX), ICH, FDA provide standards for these parameters and their acceptable thresholds. Below, each critical parameter is defined, its importance explained, how it is measured, typical values in anti-doping/forensic contexts, and illustrative examples.

7.5.1 *Selectivity (Specificity)*

Selectivity (or specificity) is the ability of a method to distinguish and quantify the analyte of interest in the presence of other substances in the sample matrix endogenous compounds, metabolites, co-administered drugs, adulterants, etc. [37, 40]. In doping control, this is vital because samples (urine, plasma, hair) contain many compounds that might interfere with detection and quantitation.

Evaluation Procedure

- Analyze multiple blank matrix samples (urine, plasma, hair) from different individuals or sources.
- Spike samples separately with analyte and internal standard.
- Examine chromatograms/mass spectrometric transitions for interfering peaks at analyte retention times or transitions.

Acceptance Criteria

Usually blanks should show no peak (or signal below a threshold, e.g., $\leq$20% of the response at LLOQ for analyte, $\leq$5% for internal standard) at analyte's retention time.

7.5.2 *Linearity and Range*

Linearity refers to the ability of the method to give test results proportional to the concentration of analyte over a given range [28, 29]. The range is the interval between the lowest and highest concentrations that can be accurately measured. Appropriate linear calibration ensures quantification is reliable.

Evaluation Procedure

- Calibration standards should be prepared over several concentration levels, typically including at least 5–8 non-zero levels in addition to a blank.
- Internal standard normalization should be applied if applicable to correct for variations in sample preparation and instrument response.
- The analytical response, such as the peak area ratio of the analyte to the internal standard, should be plotted against the known concentrations.
- The goodness of fit of the calibration curve should be evaluated using appropriate criteria, such as the coefficient of determination (r or r2), examination of residuals, and consideration of suitable weighting factors (e.g., 1/x or 1/ × 2).

Acceptance Criteria

Often r2 $\geq$ 0.99; residuals across calibration should fall within $\pm$15% ($\pm$20% at LLOQ).

7.5.3 *Limits of Detection (LOD) and Limit of Quantitation (LOQ)*

LOD is the lowest concentration of analyte that can be reliably distinguished from background noise (but not necessarily quantified with acceptable precision) [28, 29]. LOQ is the lowest concentration that can be quantified with defined accuracy and precision. In doping control, LOD and LOQ are critical for detecting trace amounts, often in prohibited substances lists, with very low thresholds.

Evaluation Procedure

- The signal-to-noise ratio method can be used, where the limit of detection (LOD) is defined as a signal-to-noise ratio of at least 3, and the limit of quantification (LOQ) is defined as a signal-to-noise ratio of at least 10.
- Alternatively, LOD and LOQ can be calculated based on the standard deviation of the blank or low-level standards and the slope of the calibration curve, using the equations LOD = 3.3 × σ/slope and LOQ = 10 × σ/slope.
- It is also necessary to demonstrate that the method achieves acceptable accuracy and precision at the proposed LOQ.

Acceptance Criteria

LOQ precision and accuracy generally $\leq$20% at LOQ; for higher levels $\leq$15%.

7.5.4 Recovery (Extraction Efficiency)

Recovery is a measure of how much analyte is retrieved from the sample matrix after sample preparation, relative to amount spiked. Consistent, high recovery ensures sensitivity and accuracy [37, 41].

Evaluation Procedure

- Blank matrix samples should be spiked with a known concentration of the analyte prior to extraction (pre-extraction spike) and also after extraction (post-extraction spike).
- The signals obtained from the pre-extraction and post-extraction spiked samples should then be compared to determine the recovery of the analyte.

Recovery = (Pre-extraction Spike/Post-Extraction Spike) x 100%.

Acceptance Criteria

Recoveries vary depending on analyte and matrix, but should be reproducible across QC levels with variability ≤15% RSD and bias within ±15% of nominal (except LOQ level where ±20%).

7.5.5 Matrix Effects

Matrix effect refers to alteration (suppression or enhancement) of analyte signal due to co-eluting matrix components affecting ionization in MS [42, 43]. In LC–MS or GC–MS, matrix effects can cause bias and reduce accuracy. It is especially significant in forensic/anti-doping because matrices are complex (urine, plasma, hair) and analytes often at trace levels.

Evaluation Procedure

- The analyte response in a pure solvent should be compared with the response of the analyte spiked into a blank matrix extract (post-extraction).
- The matrix factor (MF) is calculated as the ratio of the response in the matrix to the response in the solvent, expressed as

Matrix Factor (MF) = response in Matrix/Response in Solvent

- The percent suppression or enhancement is determined using the equation

Percent Suppression or Enhancement = (MF-1) x 100%

- Multiple lots or sources of the matrix should be used to evaluate variation in the matrix effect.

A negative percent matrix effect indicates signal suppression, where the matrix reduces the analyte response. A positive percent matrix effect indicates signal enhancement, where the matrix increases the analyte response.

Acceptance Criteria

- The matrix effect should be consistent across different matrices or lots, with variability generally not exceeding 15%.
- If significant matrix effects are observed, internal standards particularly isotopically labeled analogs should be used, and additional sample cleanup may be required.

7.5.6 Accuracy and Precision

Accuracy (sometimes called trueness or bias) is the closeness of measured values to the true value or a known reference standard [37, 44]. Precision relates to agreement between replicate measurements; includes intra-run (within a run/day) and inter-run (between days, analysts, labs) precision. Accurate and precise measurements are fundamental; a method that is precise but inaccurate will systematically misreport; accurate but imprecise gives unreliable variability.

Evaluation Procedure

- Quality control (QC) samples should be prepared at several concentration levels, including low, medium, and high.
- For precision assessment, replicate measurements (typically $n = 5$–6 or more) should be performed within a single run (intra-day) and over multiple runs or days (inter-day). The relative standard deviation (RSD) is then calculated using the equation:

Precision (%) = (Standard Deviation/Mean) x 100

- For accuracy assessment, known spiked samples are measured and the obtained concentrations are compared to the nominal concentrations. Accuracy is calculated using the equation:

Accuracy % = (Measured Value/True Value) x 100

Acceptance Criteria

Typically, the method should achieve accuracy within ±15% bias and precision with a relative standard deviation (RSD) of ≤15% at concentration levels above the limit of quantification (LOQ). At the LOQ, slightly more lenient criteria are acceptable, with accuracy within ±20% and corresponding precision.

7.5.7 *Stability*

Stability refers to whether analytes (and standards, internal standards) remain unchanged under various conditions (storage, freeze–thaw, processed sample stability, autosampler stability) [37, 45]. For anti-doping, it is critical since samples may be stored long time, shipped, subjected to temperature variation, etc.

Evaluation Procedure

- *Short-Term Stability*: Short-term stability is assessed by storing samples at room temperature for a defined number of hours and then analyzing them to determine any changes in analyte concentration.
- *Long-Term Stability:* Long-term stability is evaluated by storing samples in a freezer, typically at –20 °C or lower, for days or months, followed by analysis to confirm that the analyte remains stable.
- *Freeze–Thaw:* Freeze–thaw stability is tested by subjecting samples to multiple cycles of freezing and thawing and measuring the analyte to ensure it is not degraded.
- *Processed Sample Stability*: Processed sample stability is determined by keeping samples in the autosampler for a defined period and monitoring the analyte response over time.

7.5.8 *Carry-Over, Dilution Integrity, Ruggedness/Robustness*

Carry-Over: Carryover is the phenomenon in which analyte from a high-concentration sample affects subsequent analyses of blanks or low-concentration samples. This is particularly important in doping control, as it can lead to false positives or artificially elevated results [38]. Carryover is assessed by injecting a blank sample immediately after the highest calibrator and evaluating the blank for any detectable analyte response.

Dilution Integrity: When samples exceed the upper limit of quantification (ULOQ), they can be diluted to bring them within the calibration range. The method must demonstrate that, after dilution, the analyte can still be quantified accurately and precisely [38].

Ruggedness/Robustness: Ruggedness, or robustness, refers to the sensitivity of a method to small, deliberate variations, such as minor changes in temperature, pH, reagent lot, instrument, or operator. Evaluating ruggedness confirms that the method remains stable and reliable under normal operational conditions [38].

7.5.9 Implementation of Validation in Forensic and Anti-doping Contexts

Guidelines and Standards

- SWGTOX has published detailed guidance for forensic toxicology method validation that includes many of the parameters above.
- ISO/IEC 17,025 (requirements for testing and calibration labs) demands validated methods.
- For doping control, organizations like WADA (World Anti-Doping Agency) also require validated methods. (Though specific WADA technical documents often reference method performance criteria related to selectivity, sensitivity, etc.)

Matrices: Validation must be done in the same matrix as intended sample (urine, plasma, hair). Hair particularly presents special challenges (external contamination, low analyte concentrations, incorporation kinetics, etc.).

Use of Internal Standards: To compensate for matrix effects, analyte loss during extraction, variation in instrument response, stable isotope-labeled internal standards are highly advisable.

Low-Level Analyte Detection: Many doping substances are present at trace levels; LOQ must be low enough to detect them in real samples above thresholds used in regulations.

7.5.10 Challenges and Best Practices

- Matrix Variability: Different donors, dietary influences, disease states can affect matrix composition; validation should cover several lots/sources.
- Analyte Stability vs Storage/Shipping Conditions: In doping control, chain of custody may take weeks; need to demonstrate analyte stability over relevant time periods and conditions.
- Derivatization and Non-polar Compounds: For non-polar/poorly ionizable compounds (e.g., certain steroids), derivatization may be required especially for GC–MS. Validation must include the derivatization step (its reproducibility, effect on accuracy, precision).
- Qualitative vs Quantitative: Some tests are qualitative and only determine the presence or absence of an analyte, while others require accurate quantification. The validation approach differs accordingly: qualitative methods focus on parameters such as selectivity and limit of detection, whereas quantitative methods require a full suite of validation parameters, including precision, accuracy, and linearity.

The overall performance characteristics of an analytical method are evaluated through multiple validation parameters to ensure accuracy, reliability, and reproducibility under defined experimental conditions. Table 7.10 provides a consolidated summary of the principal validation parameters such as selectivity, linearity, limits of detection and quantification (LOD and LOQ), recovery, matrix effect, accuracy, precision, and stability along with their typical acceptance criteria as outlined by international validation guidelines. These parameters collectively form the foundation for establishing the suitability of analytical methods used in forensic toxicology and anti-doping laboratories.

Validation is not a one-size-fits-all procedure; it must be customized according to analyte class, biological matrix, instrument (LC–MS, GC–MS), detection sensitivity, and legal/regulatory thresholds. A method that passes all validation parameters under realistic sample and operational conditions is necessary to produce results that are accurate, reproducible, and defensible in forensic or anti-doping settings. Key practices include using multiple matrix lots, evaluating stability over relevant times and conditions, proper use of internal standards, ensuring LOQ meets regulatory thresholds, and verifying that derivatization or extraction methods do not introduce bias or inconsistency. Adhering to established guidelines (SWGTOX, ICH, WADA, etc.) ensures quality and comparability. Ultimately, robustly validated methods underlie

Table 7.10 Summary of key validation parameters and typical acceptance criteria used in forensic and anti-doping analytical methods

Parameter	What it measures	Typical acceptance criteria in anti-doping/forensic methods
Selectivity/specificity	No interference at analyte's signal in blank matrix	Blank $\leq$20% of LLOQ response for analyte; IS response $\leq$5%
Linearity	Proportionality of response vs concentration	$r^2 \geq 0.99$; residuals within ±15% (±20% at LOQ)
LOD/LOQ	Detection vs quantification limits	LOD S/N $\geq$ 3; LOQ S/N $\geq$ 10; LOQ precision/accuracy $\leq$20%
Recovery	Extraction efficiency	Consistent across QC levels; RSD $\leq$15%
Matrix effect	Ion suppression or enhancement	Variation $\leq$15%; use IS
Accuracy and precision	Bias and reproducibility	Bias within ±15%; precision (RSD) $\leq$15% ($\leq$20% at LOQ)
Stability	Analyte holds value under storage, etc	± 15% deviation under tested conditions
Carry-over, dilution integrity, ruggedness	Method robustness under operational conditions	No carry-over at ULOQ; diluted samples accurate; small changes tolerated

the credibility of forensic and doping control laboratories, protecting both fairness and justice.

7.6 Summary

The comparative evaluation of extraction methodologies highlights the pivotal role of selecting an appropriate sample preparation strategy to ensure optimal analytical performance in doping control investigations. SPE demonstrated broad applicability across diverse drug classes and biological matrices, whereas LLE proved particularly effective for lipophilic compounds in plasma samples. The incorporation of hair extraction techniques further enhances the detection window, enabling retrospective assessment of long-term substance use. Additionally, derivatization remains a crucial step in facilitating the GC–MS detection of thermally stable yet poorly ionizable compounds.

By integrating matrix-specific and compound-dependent extraction protocols, this chapter establishes a comprehensive framework for achieving accurate, sensitive, and reproducible analytical outcomes in forensic and anti-doping laboratories. Furthermore, critical validation parameters including selectivity, sensitivity, linearity, accuracy, precision, recovery, and matrix effects were discussed in alignment with international standards such as SWGTOX, ISO/IEC 17,025, and WADA technical guidelines, ensuring methodological reliability and regulatory compliance.

References

1. Mali N, et al. A review on biological matrices and analytical methods used in drug of abuse analysis. J Appl Pharm Sci. 2011;1(6):58–65.
2. Thevis M, Kuuranne T, Geyer H. Annual banned-substance review: analytical approaches in human sports drug testing. Drug Test Anal. 2019;11(1):8–26. https://doi.org/10.1002/dta.2549.
3. Jones S, McGowan C, Boyle S, Ke Y, Chan CHM, Hwang H. An overview of sample preparation in forensic toxicology. Wiley Interdiscip Rev Forensic Sci. 2022;4(2): e1436.
4. Olariu RI, Vione D, Grinberg N, Arsene C. Sample preparation for trace analysis by chromatographic methods. J Liq Chromatogr Relat Technol. 2010;33(9–12):1174–207. https://doi.org/10.1080/10826076.2010.484371.
5. Andersen DW, Linnet K. Screening for anabolic steroids in urine of forensic cases using fully automated solid phase extraction and LC-MS-MS. J Anal Toxicol. 2014;38(9):637–44. https://doi.org/10.1093/jat/bku098.
6. Ohnuma K, Hirano-Kodaira M, Bannai M, et al. A broad-spectrum peptide screening method using an optimized solid-phase extraction and liquid chromatography-high-field asymmetric ion mobility spectrometry-mass spectrometry for doping control in equine urine. Anal Methods. 2024;16(47):8250–67. https://doi.org/10.1039/d4ay01477d.
7. Kwok WH, Choi TLS, Tsoi YYK, Leung GNW, Wan TSM. Screening of over 100 drugs in horse urine using automated on-line solid-phase extraction coupled to liquid chromatography-high resolution mass spectrometry for doping control. J Chromatogr A. 2017;1490:89–101. https://doi.org/10.1016/j.chroma.2017.02.020.

8. Frederick F, Cantwell ML. Chapter 11: liquid-liquid extraction. In: Comprehensive Analytical Chemistry, Vol. 37, Elsevier, pp. 297–340 (2002). https://doi.org/10.1016/S0166-526X(02)80048-4
9. Lepoittevin M, Blancart-Remaury Q, Kerforne T, et al. Comparison between 5 extraction methods in either plasma or serum to determine the optimal extraction and matrix combination for human metabolomics. Cell Mol Biol Lett. 2023;28(1):43. https://doi.org/10.1186/s11658-023-00452-x.
10. Temerdashev A, Nesterenko P, Dmitrieva E, et al. GC-MS/MS determination of steroid hormones in urine using solid-phase derivatization as an alternative to conventional methods. Molecules. 2022;27(18):5796. https://doi.org/10.3390/molecules27185796.
11. Tamama K. Dilute and shoot approach for toxicology testing. Front Chem. 2023;11:1278313. https://doi.org/10.3389/fchem.2023.1278313.
12. Vandenberghe-Dürr S, Harr T, Vandenberghe F. Antihistamines: an ongoing narrative. Rev Med Suisse. 2024;20(868):711–9.
13. Subhahar MB, Karakka Kal AK, Philip M, et al. Detection and identification of ACP-105 and its metabolites in equine urine using LC/MS/MS after oral administration. Drug Test Anal. 2021;13:299–317. https://doi.org/10.1002/dta.2918.
14. Philip M, Karakka Kal AK, Subhahar MB, et al. Characterization of growth hormone secretagogue small molecule ibutamoren (MK-0677) and its possible metabolites in thoroughbred horses for doping control. Rapid Commun Mass Spectrom. 2022;36(18): e9337. https://doi.org/10.1002/rcm.9337.
15. Philip M, Karakka Kal AK, Subhahar MB, et al. Hydroxy levamisole and its phase II conjugates as potential indicators of levamisole doping in thoroughbred horses. Rapid Commun Mass Spectrom. 2023;37(3): e9430. https://doi.org/10.1002/rcm.9430.
16. Blower PR. Non-steroidal anti-inflammatory drugs. Br J Rheumatol. 1993;32(Suppl 4):35–8. https://doi.org/10.1093/rheumatology/32.suppl_4.35.
17. Johnson AG, Quinn DI, Day RO. Non-steroidal anti-inflammatory drugs. Med J Aust. 1995;163(3):155–8. https://doi.org/10.5694/j.1326-5377.1995.tb127972.x.
18. Subhahar MB, Singh J, Albert PH, Kadry AM. Pharmacokinetics, metabolism and excretion of celecoxib, a selective cyclooxygenase-2 inhibitor, in horses. J Vet Pharmacol Ther. 2019;42:518–24. https://doi.org/10.1111/jvp.12757.
19. Mathew B, Philip M, Perwad Z, et al. Identification of hypoxia-inducible factor (HIF) stabilizer roxadustat and its possible metabolites in thoroughbred horses for doping control. Drug Test Anal. 2021;13:1203–15. https://doi.org/10.1002/dta.3014.
20. Kapugi M, Cunningham K. Corticosteroids. Orthop Nurs. 2019;38(5):336–9. https://doi.org/10.1097/NOR.0000000000000595.
21. De Bock M, Sienaert P. Corticosteroids and mania: a systematic review. World J Biol Psychiatry. 2024;25(3):161–74. https://doi.org/10.1080/15622975.2024.2312572.
22. Karatt TK, Nalakath J, Perwad Z, et al. Mass spectrometric method for distinguishing isomers of dexamethasone via fragment mass ratio: an HRMS approach. J Mass Spectrom. 2018;53:1046–58. https://doi.org/10.1002/jms.4279.
23. Karakka Kal AK, Karatt TK, Nalakath J, et al. Simultaneous analysis of 44 frequently abused corticosteroid drugs using polysaccharide-based chiral column-HRMS approach. Anal Sci Adv. 2021;2:427–39. https://doi.org/10.1002/ansa.202000166.
24. Kunhamu Karatt T, Sathiq MA, Laya S, et al. Is 9β-dehydrohalogenation of betamethasone and dexamethasone hindering the detection of banned co-eluting meprednisone? A reverse-phase chiral liquid chromatography high-resolution mass spectrometry approach. Steroids. 2020;155: 108572. https://doi.org/10.1016/j.steroids.2019.108572.
25. Bond P, Smit DL, de Ronde W. Anabolic-androgenic steroids: how do they work and what are the risks? Front Endocrinol (Lausanne). 2022;13:1059473. https://doi.org/10.3389/fendo.2022.1059473
26. Reyes-Vallejo L. Current use and abuse of anabolic steroids. Actas Urol Esp (Engl Ed). 2020;44(5):309–13. https://doi.org/10.1016/j.acuro.2019.10.011.

27. Fosgerau K, Hoffmann T. Peptide therapeutics: current status and future directions. Drug Discov Today. 2015;20(1):122–8. https://doi.org/10.1016/j.drudis.2014.10.003.
28. Berillo D, Yeskendir A, Zharkinbekov Z, et al. Peptide-based drug delivery systems Medicina (Kaunas). 2021;57(11):1209. https://doi.org/10.3390/medicina57111209.
29. Duarte CD, Greferath R, Nicolau C, et al. Myo-inositol trispyrophosphate: a novel allosteric effector of hemoglobin with high permeation selectivity across the red blood cell plasma membrane. ChemBioChem. 2010;11(18):2543–8. https://doi.org/10.1002/cbic.201000499.
30. Dan W, Gao J, Qi X, et al. Antibacterial quaternary ammonium agents: chemical diversity and biological mechanism. Eur J Med Chem. 2022;243: 114765. https://doi.org/10.1016/j.ejmech.2022.114765.
31. McClung MR. Bisphosphonates. Endocrinol Metab Clin North Am. 2003;32(1):253–71. https://doi.org/10.1016/s0889-8529(02)00079-8.
32. Vepsäläinen JJ. Bisphosphonate prodrugs. Curr Med Chem. 2002;9(12):1201–8. https://doi.org/10.2174/0929867023369998.
33. Qallaf MA. Different extraction procedures for multiple drugs abuse in human hair: case study. Chem Pharm Res. 2022;4(1):1–6.
34. Zhang HH, Loh JYH, Moy HY, et al. An improved GC-MS/MS method for a fast multidrug analysis in hair. Drug Test Anal. 2025;17(8):1344–56. https://doi.org/10.1002/dta.3840.
35. Karatt TK, Sathiq MA, Laya S, et al. In-depth metabolic study of nonsteroidal selective androgen receptor modulator GSK2881078 in thoroughbred horses and horse liver microsomes for doping control. Drug Test Anal. 2023;15(7):757–68. https://doi.org/10.1002/dta.3469.
36. Ramoo B, Funke M, Frazee C, et al. Comprehensive urine drug screen by gas chromatography/mass spectrometry (GC/MS). Methods Mol Biol. 2016;1383:125–31. https://doi.org/10.1007/978-1-4939-3252-8_15.
37. Woźniak MK, Banaszkiewicz L, Wiergowski M, et al. Development and validation of a GC–MS/MS method for the determination of 11 amphetamines and 34 synthetic cathinones in whole blood. Forensic Toxicol. 2020;38:42–58. https://doi.org/10.1007/s11419-019-00485-y.
38. Arantes ACF, da Cunha KF, Cardoso MS, et al. Development and validation of quantitative analytical method for 50 drugs of antidepressants, benzodiazepines and opioids in oral fluid samples by liquid chromatography–tandem mass spectrometry. Forensic Toxicol. 2021;39:179–97. https://doi.org/10.1007/s11419-020-00561-8.
39. Karatt TK, Sathiq MA, Laya S, et al. An extensive screening method for the identification and quantitation of ecdysteroids in equine urine and plasma using liquid chromatography coupled with mass spectrometry. Rapid Commun Mass Spectrom. 2023;37(18): e9611. https://doi.org/10.1002/rcm.9611.
40. Ajeebsanu M, Koshy S, Karakka Kal AK, et al. Suzetrigine in equestrian sports: optimized extraction and LC-HRMS detection strategies. Rapid Commun Mass Spectrom. 2025;39(23): e10135. https://doi.org/10.1002/rcm.10135.
41. Karakka Kal AK, Nalakath J, Karatt TK, et al. Development and validation of a chiral LC-MS method for the enantiomeric resolution of (+) and (-)-medetomidine in equine plasma by using polysaccharide-based chiral stationary phases. Chirality. 2020;32:314–23. https://doi.org/10.1002/chir.23166.
42. Karakka Kal AK, Subhahar M, Philip M, et al. Equine metabolism of voxelotor and its impact on hematological indices: a doping control perspective. Drug Test Anal. 2025;1–14. https://doi.org/10.1002/dta.3957
43. Stelmaszczyk P, Gacek E, Wietecha-Posłuszny R. Optimized and validated DBS/MAE/LC–MS method for rapid determination of date-rape drugs and cocaine in human blood samples—a new tool in forensic analysis. Separations. 2021;8(12):249. https://doi.org/10.3390/separations8120249.
44. Singh B, Sharma S, Dubey N, et al. Bioanalytical method development and validation of alimemazine in human plasma by LC-MS/MS and its application in bioequivalence studies. J Pharm Bioallied Sci. 2013;5(4):257–64. https://doi.org/10.4103/0975-7406.120068.
45. Karatt TK, Ajeebsanu MM, Karakka Kal AK, et al. Electrospray ionization mass spectrometry adduct formation by mobile phase additives: a case study using nitrile functional

groups containing selective androgen receptor modulators. Rapid Commun Mass Spectrom. 2023;37(14): e9530. https://doi.org/10.1002/rcm.9530.

MIX
Papier aus verantwortungsvollen Quellen
Paper from responsible sources
FSC® C105338

If you have any concerns about our products,
you can contact us on
ProductSafety@springernature.com

In case Publisher is established outside the EU,
the EU authorized representative is:
Springer Nature Customer Service Center GmbH
Europaplatz 3, 69115 Heidelberg, Germany

Printed by Libri Plureos GmbH
in Hamburg, Germany